Science and politics

303.483 BOG
SCIENTIFIC SECTION
HOUSE OF COMMONS LIBRARY

LOCATION
£10.95
Research Sc
AUTHOR
BOGDANOR
DATE of CATALOGUING
28 FEB 85
HOUSE OF COMMONS LIBRARY

HOUSE OF COMMONS
LIBRARY
DISPOSED OF BY
AUTHORITY

TO BE
DISPOSED
BY
AUTHORITY

Science and politics

THE HERBERT SPENCER LECTURES 1982

Edited by

VERNON BOGDANOR
Fellow of Brasenose College, Oxford

CLARENDON PRESS · OXFORD
1984

Oxford University Press, Walton Street, Oxford OX2 6DP
London New York Toronto
Delhi Bombay Calcutta Madras Karachi
Kuala Lumpur Singapore Hong Kong Tokyo
Nairobi Dar es Salaam Cape Town
Melbourne Auckland
and associates in
Beirut Berlin Ibadan Mexico City Nicosia

Oxford is a trade mark of Oxford University Press

Published in the United States
by Oxford University Press, New York

© *University of Oxford 1984*

All rights reserved. No part of this publication may be reproduced, stored in a retrieval system, or transmitted, in any form or by any means, electronic, mechanical, photocopying, recording, or otherwise, without the prior permission of Oxford University Press

British Library Cataloguing in Publication Data
Science and politics.—(The Herbert Spencer lectures; 1982)
1. Science and state
I. Bogdanor, Vernon II. Series
500 Q125
ISBN 0-19-857605-6

Library of Congress Cataloging in Publication Data
Main entry under title:

Science and politics.

(The Herbert Spencer lectures; 1982)
Bibliography: p.
Includes index.
1. Science—Social aspects. I. Bogdanor, Vernon, 1943- . II. Series.
Q175.5.S335 1984 303.4'83 84-991
ISBN 0-19-857605-6

Typeset by DMB Typesetting, Oxford
Printed in Great Britain by
St Edmundsbury Press, Bury St Edmunds, Suffolk

Preface

Herbert Spencer, like so many of the great Victorians, straddled the conventional boundaries which demarcate academic disciplines. As knowledgeable about science as about government and politics, he devoted his life to the promulgation of a systematic science of society, an undertaking which few contemporary sociologists would regard as at all feasible today. Living in an age of hitherto unparalleled scientific advance, he nevertheless did not foresee—how could he have done?—the complex ethical questions which would be posed by the consequences of scientific discoveries. To the question of the proper relationship between science and politics, he offered but a short answer—each should remain within its own respective sphere. The scientist had no business dabbling in political affairs, while the state should in turn avoid any interference with science, restricting itself to its most basic functions—the defence of the realm, the preservation of law and order, and the upholding of liberty.

Such an answer, as all the contributors to this book agree, is hardly possible today. The intrusion of politics into areas of life previously regarded as non-political is indeed one of the most momentous changes which the twentieth-century has wrought—and the scientist has, no more than any other citizen, been able to escape the burden of choice and responsibility which this lays upon him. These lectures seek to clarify the nature of that burden, and to illustrate those fearful conflicts of responsibility which accompany the awesome power of twentieth-century science.

Sir Alan Cottrell provides a synoptic account of the moral problems which face both scientists and governments in areas such as nuclear physics, ecology and the medical sciences. Raymond Aron shows how the responsibilities which scientists believe they have towards their own country and way of life have led to 'the permanent mobilization of science in the service of armaments since 1945'. Don K. Price, like Aron a political scientist, shows that not only have government and politics invaded the realm of science, but science has also profoundly affected the operation of political institutions. Margaret Gowing illustrates the influence of science on political revolutions. The Bolshevik revolution itself—perhaps the most significant political event of the twentieth century—was carried out by men who believed, sincerely if misguidedly, that their ideology enjoyed the endorsement of a scientific guarantee, being in accordance with methods which had already ensured triumphant advances in the natural sciences. The physicist H. Maier-Leibnitz, and the pharmacologist Sir William Paton, both show how the ignorance of laymen and governments about science lead them to misunderstand and misconstrue the seriousness with which scientists take their ethical responsibilities. For it is the sharpening of conflict in the political world, not the irresponsibility of the scientist, which so threatens the future of humanity. In his discussion of the arms race, Aron shows that it is a product not so much of science, but rather of fundamental conflicts between competing ideologies and national interests—a product, in short, of human nature. The same factors which have, from the beginning of time, generated human conflict are still in operation, but the power which science has unleashed makes it more imperative than it has ever been that man's destructive impulses be controlled. Indeed, these lectures pose a challenge as much to governments and students of society as they do to scientists. For it is the task of both sociologist and political scientist to seek to understand the workings of political society as successfully as twentieth-century scientists have advanced our understanding of the natural world. The student of society must be imbued with a sense of urgency as he attempts to answer questions which, for Herbert Spencer, were no more than a speck on the intellectual horizon.

There is probably no one in the twentieth century who grappled more seriously, from the standpoint of political science and sociology, with the ethical problems with which these lectures deal than Professor Raymond Aron. It was therefore with deep sadness that we learnt of his death in October 1983. Aron's whole personality and intellectual outlook, that of a liberal sceptic in an age of competing faiths, offers a striking illustration of Herbert Spencer's dictum that the man of 'higher type . . . must be content with greatly moderated expectations, while he perseveres with undiminished efforts. He has to see how comparatively little can be done, and yet to find it worth while to do that little: so uniting philanthropic energy with philosophic calm' (*The Man versus the State*, Penguin edition, Harmondsworth, 1969, page 191).

April 1984

Vernon Bogdanor
Brasenose College, Oxford

The Clarendon Press is especially grateful to Vernon Bogdanor, who kindly agreed to edit this volume at a late stage when the editor originally appointed was unable to complete the task.

Contents

1 *The sorcerer's apprentice*

ALAN COTTRELL

Jesus College, Cambridge

Lower than the angels

Among the foundations of mankind's ancestral beliefs and general precepts for living, which largely continue to the present day, is a simple and basic but largely unrecognized principle: that mankind is weak, puny, and insignificant, in relation to the forces of nature. The earliest peoples, unable to control, influence, or even comprehend the many natural disasters and personal tragedies which so often devastated them, found comfort in a belief that real power belongs to supernatural beings whose favours might be won by assiduous devotion and sacrificial offerings. From this grew later beliefs in, on the one hand, demons, witches, and spirits, beliefs which to some extent can still be seen in today's fascinations with poltergeists, horoscopes, and occult practices; and, on the other hand, faiths in great religions which, by encouraging prayers to God for good harvests, safe voyages, deliverance from sickness, and for other benefits, fully endorsed traditional ideas of human helplessness and divine powerfulness.

But the principle extends to many other human beliefs and practices. The numerical insignificance of early populations, scattered over great empty tracts of the earth, found its echo in the Biblical imperative 'Be fruitful and multiply' out of which has grown one of our most deeply held human 'rights', the right of parents to raise large families as they please. The weakness of human effort in wresting a living from a recalcitrant earth, again echoed in a biblical saying, 'In the sweat of thy face shalt thou eat bread', has led to the work ethic, to a belief that it is our destiny to fill the hours of the day with toil and labour, a belief in the virtues of a busy life of low productivity. We have also supposed that the worst that we could do to the earth itself, through plundering its natural resources and polluting it with wastes, is insignificant because of the giant size of the earth and its enormous powers of recovery from such insults. Economists, whether classical or Marxist, have, almost without exception, simply taken the earth for granted, as an inexhaustible supplier of free goods and a bottomless rubbish dump, an assumption which could only be valid for a powerless and physically insignificant mankind. Even war has been seen, at least in past centuries, as a fairly small-scale activity by professional armies, without too much harmful effect on mankind and society generally; and in medicine the traditional ethic that a physician must do everything in his power to prolong life as much as possible has rested on the chastening knowledge that, all too often, this power was very small indeed.

That all these beliefs, customs and practices rest upon the general assumption of mankind's powerlessness is not surprising, since they have grown up over many centuries, during most of which the assumption was clearly justified. The world's population in biblical times was probably less than 300 million. Famine and plague made sure that it would not rise far above this figure for many centuries to come. There was no antiseptic medicine or public health. Most working people had only their own limbs to help them; the occasional one had an ox, mule or horse. Goods were sparse and made largely from natural products, hand-crafted in cottage industries. Travel was difficult, slow and dangerous. Even as late as the eighteenth century, warfare was still conducted on limited battlefield sites between small standing armies using hand weapons or small cannons.

Modern science and technology have changed all this. Today, we have vast physical powers. But we still run our lives and societies according to the old rules, which only make sense on the now false assumption of the comparative powerlessness of mankind. These rules have slowly evolved over many centuries and are now deeply ingrained habits, which can hardly be uprooted. Nor is it easy to find, quickly, good replacements for them, to match the present times. This problem of the adaptation of mankind to the unfamiliar possession of real power, the problem of the 'sorcerer's apprentice', is the subject of our lecture.

The years immediately after the Second World War were a time of hope in a scientific future. People saw in science and technology an all-conquering instrument for achieving national goals. Science decreed what could be done and technology decreed how to do it. Together they provided power, the power of new ideas embodied in material strength. Admittedly, all this hope was overshadowed by the threat of the atomic bomb, but nevertheless there was an optimistic belief that science and technology could lead mankind on to such a bright future that the fearful needs for stockpiled weapons would eventually fade away. Electronics, automation and computers would provide plentiful goods for all and take the drudgery out of work. Civil nuclear energy would provide limitless electrical power. New drugs would cure our ills and new agricultural chemicals would solve our food problems. Everyone would have a motor car and take holidays abroad by cheap air travel. There would be lots of free time for all.

Of course, since then a lot of this has actually happened. But so often there has been some kind of a twist to it, some unexpected and often unwelcome deviation from those bright hopes of 30 years ago. We do have the cheap mass travel, but we also have the noise, stench, and congestion on our roads and at the airports. We have the drugs, but also the thalidomide tragedies. People with a lifetime's skills are thrown out of work by new technology which eliminates the need for their services.

All this justifies the view that we have not yet learned to make sensible use of science and its applications. But let us also remember that we generally take our blessings, including scientific blessings, for granted and are inclined to focus on the bad things just as obsessively as we ignore the good ones. Let us remember that there has been no war in Europe for a third of a century now, which is an almost miraculous feat of self-restraint in this traditionally warlike quarter of the globe. While unemployment has grown high again and many people suffer poverty, this has not reached the state of barefoot misery and cruel hunger which earlier generations suffered. And to compare conditions over a longer timescale, I would like to take you briefly back to a time before the Industrial Revolution, as described in J. H. Plumb's book.[1] Referring to the towns of England, he writes:

> The first noticeable thing about these towns would have been the stench. There was no sanitary system. . . . The houses of the poor were one or two-room hovels . . . most cellars were inhabited, not only by people but also by their pigs, fowl, sometimes even by their horses and cattle. All houses and

cellars were desperately overcrowded; ten to a room was common. . . . Disease was rampant and unchecked: smallpox, typhus, typhoid, and dysentery made death a commonplace . . . only about one child in four, born in London, survived. . . . The craftsmen and artisans worked long hours—fourteen was common—but trade was fickle and the chance of hunger and poverty threatened their lives with anxiety. . . . Below the artisans and journeymen were the mass of London's population, the hordes of labourers whose livelihood depended almost entirely on casual employment. . . .

The major cause of all that misery, poverty and disease was mankind's lack of scientific knowledge and technological power. The history of the Industrial Revolution and of the ensuing scientific improvement of society makes this quite clear, bringing us right up to the present day, when most people, in industrialized countries at least, live in houses or flats with several rooms to a family, fresh and safe drinking water on instant delivery from an internal tap, inside lavatories and bathroom, heat, lighting, and usually television for entertainment. The average working day has been halved. There are plenty of holidays and large numbers of people are able to afford and make journeys abroad, comfortably within an hour or two, journeys that would have taken Byron and Mozart days in juddering coaches. People are also protected by a National Health Service and a system of social security; and there is free state education.

All this has been brought about in 200 years, made possible by the scientific, technological and industrial developments during that period. It is an almost miraculous transformation and we ought not to forget these benefits of scientific power as we turn to the problems which it also raises.

Power and responsibility

With power should come responsibility. How well have we exercised our new powers over nature? We have certainly multiplied our populations. It took about a million years to reach the first 1000 million, worldwide, in the mid-nineteenth century. It now takes only twelve years to add another 1000 million. What mankind has mainly done with its new scientific and technological powers is to multiply the human species. There is no wisdom or responsibility in this; any animal species would have done just the same, given the same physical opportunities.

As a result of this population explosion, mankind is now irreversibly committed to the continuance of technological development. There can be no going back. The argument has nothing to do with a reluctance to give up modern luxuries. It is simply that there are so many people in the world. They could not survive without science and technology, whether this is providing a green revolution in India, a system of public health in Hongkong, an airlift of life-saving supplies to an earthquake disaster zone, or winter heating in northern Europe and America.

Because technology has been used to increase population, technology now has to race further ahead to keep abreast of the growing world populations and to try to meet demand for a better standard of living, a demand which is unquestionably justified in at least the third world countries. Mankind is now committed, through population growth, to a technological escalator.

Where will this escalator take us? Within at most a few generations from now the world's population will level out. Inescapable scientific facts will see to that, if necessary in the grimmest possible way, even if the ingenuity of mankind continues for a while to put off the day of reckoning. But in the nearer future, over the next 50 years or so, what of the effects then? The most serious will be the coming world shortage of energy and, stemming from that, a shortage of food, goods and employment. In energy needs alone, the most sober and responsible estimates point to a four-fold increase, world wide, over the next 40 years. It is very hard to see from where that could now come. I do not believe that technology will be able, for more than a few years, to keep pace with the present rate of world population growth.

Environmental effects

By contrast, I do not believe that the environmental effects of technology, world wide, will be as serious as some people have feared, although there are some considerable hazards to be faced. We must always remember the gigantic size of the earth's natural processes and the smallness of mankind—even modern, powerful, multitudinous mankind—by comparison. Sunlight pours down on earth, daily, 20000 times as much energy as all mankind uses; and a single hurricane releases energy at the rate of several million power stations.

It is necessary however to look further, into more subtle effects where a small cause might trigger off a large change. It would be quite impossible here to review all the possible hazards which come from the release of traces of poisonous substances such as lead, other heavy metals, insecticides, weedkillers, aerosols, and sulphurous gases, into the air, water, and food chains. A great deal of careful study now goes into them and many countries control the purity and safety of their environments with strict legal regulations. There is always a danger of taking too simple a view, in one direction or the other, of these possible environmental hazards, for the effects which have to be considered are very numerous, often quite subtle and indirect, and the policy decisions about them call for difficult judgements in balancing benefit against hazard. For example, widespread DDT seriously harms many creatures, particularly predatory birds, yet its benefits to people in poor tropical countries are immense.

The greatest environmental dangers are coming at present, not from our latest and most sophisticated science, but from older and simpler scientific developments which are acting indirectly, through the pressure of growth in human populations and the physical powers available to modern man. Intensive cattle grazing and agriculture are causing the deserts to expand at alarming rates in North Africa and some other parts of the world. Many tropical forests are being cut down at a huge rate by overpopulated subsistence farmers, desperate for cropland and firewood, and this is leading to catastrophic losses of topsoil. Acid rain due to fossil-fuel burning in industrial countries is believed to be killing forests and lake fish in northern Europe and America. Large parts of the Baltic Sea, the Mediterranean, and the great lakes in America and the Soviet Union have become seriously polluted; and sea and coastal pollution by oil spills has now become commonplace. Many of the world's traditional sea fisheries are now seriously depleted through over-fishing. Looking to the future, there are ambitious large-scale schemes such as the Soviet Union's proposed plan to redirect Siberian rivers from the Arctic Ocean to the arid farmlands of central Asia. By so increasing the salinity and lowering the freezing point of this ocean, such a change could trigger vast and unpredictable climatic changes over the northern hemisphere, because its present frozen state is unstable and might be upset by a small provocation. To bring about, inadvertently, such a vast and probably irreversible change would indeed justify the label of 'sorcerer's apprentice'.

The legacy of nuclear physics

In 1947 Robert Oppenheimer said that 'the physicists have known sin; and this is a knowledge which they cannot lose'. The power which he and his colleagues in nuclear physics gave to the world a few years earlier had indeed been unleashed in a most terrible form, in the atomic bombing of Hiroshima and Nagasaki. Although this brought an immediate end to the Second World War and perhaps prevented the even greater havoc of an attempted invasion of Japan, nevertheless I think that most of us have wished that the nuclear physicists had not made those terrifying discoveries; that the knowledge which we cannot now lose might never have been won in the first place. But we cannot go back. We must look to the present and the future.

What we find here is the *technological arms race*, mainly between the United States and the Soviet Union, in which the defence scientists strive ceaselessly to keep ahead of the other side in devising new kinds of nuclear warheads, new forms of delivery missiles, and new methods of destroying those missiles. There is truth in the remark that each new weapon is already obsolete by the time that it goes into production. Each side remains fearful of being taken by surprise by the other; each boosts its arguments with a 'worst case' analysis of the potentialities of the other; and all this drives both into an endless, restless and limitless exploration of new military possibilities at the frontiers of scientific knowledge. Galbraith has described this as the *technological trap*: 'Each power develops the weapons which make obsolete those of the other. Anticipating this, each strives to develop those that protect it from obsolescence and provide an advantage instead. The resulting interaction has a technological dynamic of its own.' At first this came about through radar, jet engines and the atomic bomb. Then—and in addition to the ever-continuing development of these—were added the hydrogen bomb, space rockets, and satellites. Now, still further on, we are seeing the exploration of the potentialities of lasers and particle beams for shooting down missiles and satellites. Real science and fictional science begin to merge, in a kind of *Star Wars* scenario.

This escalator fails even in its primary purpose since it gives both sides not greater security, but the reverse—more insecurity, more defence instability, more fear of being caught by surprise. How can we step off it? The real solution can only be a political one, a growth in each side of confidence in the trustworthiness of the other side. In the face of such a distant goal the remaining possibilities are few, but they ought nevertheless to be strenuously pursued. One which we ought to press for hard is the general extension of test ban treaties, since new weapons are less likely to be developed or used if they can never be tested and proven. Another possibility would be to open up, perhaps through a revitalized United Nations, a standing conference between defence scientists from the two sides about the military potentialities of scientific knowledge. If successful, this would remove some of the fears of each side that the other one was leaping ahead in defence science. There are precedents for precisely this kind of exchange of knowledge and views, in the 1955 Atoms for Peace Conference, and in discussions on the geophysical problems of detecting and identifying distant underground nuclear test explosions. Of course such conferences would suffer initially from infinite caution and suspicion, but nevertheless I believe that, with persistence, they could eventually lead to genuine exchanges of knowledge and views. A scientific conference, with its tradition of disclosure and free debate, and where the participants are anxious to show how successful they are scientifically, is just about the most difficult place to keep secrets.

Brave new world

Let us turn now to the medical sciences. Here the predicament of scientific man is only too obvious. The brilliance of biological discoveries and the virtuosity of medical techniques has given him godlike powers over life and death, even to create new forms of life. But a lesson we are learning slowly from this is how exceedingly difficult it is to use these biomedical powers with godlike wisdom. In the old days of fairly powerless medicine, the ethical problem was quite straightforward. The physician simply applied all his lifesaving skill and knowledge, such as it was, but nature nevertheless usually took its course relentlessly. Today, by contrast, we have most powerful life-support systems and intensive-care units in hospitals. These triumphs of medical science can perform miracles by really altering the course of nature, even staying the hand of mortality itself. As ordinary human beings we are profoundly grateful to them, for there are many of us today who, through them, are enjoying second lives. But these new godlike medical powers have brought vast new responsibilities and new ethical and legal problems of the utmost difficulty. Where is wisdom, or even decent humanity, when life-support machines have kept a last wisp of life flickering in an all-but-dead political leader, stretched out for week upon week until it becomes an obscenity? And where is legality, or even plain morality, when a doctor or nurse is publicly vilified for not doing the same for a cruelly handicapped infant or a long-suffering invalid? The grand principles of a civilized, caring society reveal their inadequacy in the face of such an issue: when we fudge our responsibility and leave the problem, with no clear guidance, to the medical staffs. If that is the best we can do—and it may be the wisest course to take—then at least we ought to show a more humble attitude to the awesome decisions that doctors and nurses actually take.

But ought we to go on adding to biomedical knowledge, in directions that could lead to still more fearful consequences? This is the dilemma raised by recent researches. It has appeared in several different areas of scientific enquiry; for example in genetics, in the question of a connection between race and intelligence; and in neurophysiology, through the possibility of changing personality. But outstandingly it has appeared in two fields, embryology and genetic engineering.

Aldous Huxley's *Brave New World* shocked people with the idea of test-tube babies, but what was then science fiction has now become fact. Eggs are now removed from mothers-to-be, fertilized in a test tube and then briefly incubated into embryos before being returned to the womb. This has been done several times now, particularly in this country and Australia, and has proved a blessing to those parents who would otherwise be infertile. But again it has raised deep ethical and legal questions which seem to call for superhuman wisdom and understanding. More embryos are gathered in the test-tube than are 'needed'. Someone then has to select which is to go forward. How are they to do that? Ought they to do that? And what of the rejected ones? Are they to be let die and does that constitute killing human life? If not, are they to be deep frozen and stored for some future 'use'. If so, what use? The ethical and legal questions rise up like a forest before us, and who has the wisdom to answer them? It will be interesting to see what conclusions the public enquiry into this will reach.

Looking further ahead, there is the still more fearful possibility of *cloning*. By replacing the nucleus of a fertilized egg with the nucleus of a blood cell of the individual to be copied, an identical human being could be reproduced. This has already been done with some lower animals and the day is not far off when as many identical human beings as we wish could be replicated. Perhaps the only possible attitude to such a prospect as this is Lord Rothschild's black humour, when in 1975 he asked whether we would rather have 1000 Harold Wilsons or 1000 Marilyn Monroes. Or both.

Genetic engineering based on the technique of *recombinant DNA* takes us even beyond this. By using enzymes to unite segments of DNA from different species, entirely new kinds of genetic blueprints can be artificially constructed which, introduced into bacterial cells, can then be multiplied indefinitely to produce organisms never before seen in nature. The biological powers that this technique promises to mankind are almost beyond measure. Great benefits may come, through the ability to manufacture insulin, interferon, monoclonal antibodies, and various medicinal, veterinary and agrochemical substances. It may also become possible to develop new strains of crops that can fix their own nitrogen. Against this are the dangers, such as that of ubiquitous and previously harmless bacteria developing the ability to produce cholera or botulism, through a change in their genetic constitution.

About ten years ago fears of this kind set off a fierce debate, mainly in the United States, and the Mayor of Cambridge, Massachusetts, imposed a ban on the riskier experiments in genetic engineering, which remained in force for several months until lifted by a review board of local citizens. Today, that worry has mainly abated, partly because careful analyses have shown that the chances of a dangerous pathogenic breakout are much smaller than were previously believed, and partly because regulatory controls have been devised and accepted which strictly confine the more risky experiments to safe enclosures. But there are some lasting effects of that debate. A few leading scientists deliberately abandoned their lines of research, for fear of the consequences, and the question has been asked whether there is some science that ought not to be done, some apples which ought to be left on the tree of knowledge, unplucked.

Limits to scientific enquiry

This raises the question of whether limits ought to be set to scientific enquiry. There is of course nothing new in this. Centuries ago, Galileo's telescope was feared for the forbidden knowledge it was revealing. Darwin was similarly opposed in the last century, and to some extent is also today. There were even fears, a hundred years ago, that the steam engine and the telegraph were so disturbing to the human nervous system as to imperil civilization itself. And of course, limits have long been set and accepted on the extent to which experiments can be done on human subjects.

Even where it might be desirable, it is hardly practicable to put restraints on the human urge to seek knowledge. In the interests of preventing the spread of nuclear weapons, knowledge of the gas diffusion method of separating uranium isotopes has been kept a strict secret; but the effect of this has been simply to stimulate several countries into developing entirely new and in some respects superior methods of separating these isotopes. In pure science the difficulties are even greater. Unsolved scientific problems stand, like unclimbed mountains, as irresistible challenges to the adventurous human mind. They will always be tackled by someone, somewhere, simply because they are there. Sooner or later they get scaled.

In any case, who would hold the right to decide which scientific researches should be done or not? Would this be some national tribunal with representatives of the general public, the church, the law, the trades unions, etc.? And from where would these guardians get the foresight and wisdom to see the future consequences of these researches, when the scientists themselves, the front line troops of science, are constantly surprised by the completely unexpected turns which their investigations take. What superhuman being could have foreseen that Hertz's investigations into electromagnetic waves would give political leaders the technical means for dictatorial control of their peoples, or that Perkins' researches on dyestuffs would destroy the livelihoods of a million indigo workers in India, or that Einstein's studies of special relativity would provide the basic equation for nuclear energy?

The control of pure science is thus completely impracticable. But it is also, I submit, completely undesirable. We shall not solve the world's problems of food, health, energy, employment etc. by putting a stop to new knowledge. And if we start by curtailing scientific freedom we may end by persecuting another Galileo, martyring another Vavilov, and burning books in true Nazi style.

Thomas Jefferson once said 'there is no truth on earth that I fear to be known'. We are in danger today of losing his confidence. The question to ask is—which do we fear more: the results of scientific research or our ability to use those results wisely? The answer must surely be the second of these, in which case the right way for us must be, not to push science backward, but to pull our social, political and legal capabilities forward, to make them equal to the new challenges set by science and new powers that mankind has acquired.

Note

1. J. H. Plumb. *England in the eighteenth century*, p. 12. Penguin, Harmondsworth (1963).

2 *Science, the arms race, and arms control*

†RAYMOND ARON

In the aftermath of the French defeat of 1870, Ernest Renan, in his work, *La Réforme Intellectuelle et Morale de la France*, wrote: 'War is from now on a matter of science'. The idea is not new but from time to time, in periods of peace, it was forgotten. Military leaders make do with the weapons they have and do not ask scientists to supply better ones. Auguste Comte, with his typical mix of the profound and the bizarre, founded his thesis of the essentially peaceloving nature of modern societies on the fact that they did not use all the means of warfare that science could put at their disposal. He added that in the event that these societies would be dragged into war, the potential weapons of science would emerge. He was not wrong.

Since 1945, the connection between science and armaments has taken on a new character. First and above all, research on totally new or more efficient weapons has never ceased. What occurred between 1914 and 1918, and then again on a larger scale between 1939 and 1945 became permanent: the mobilization of science and technology in the service of armaments. In this sense, the qualitative arms race became a permanent, indeed normal situation. Before 1914, the arms race was essentially quantitative—a question of the number of divisions and warships. Three-year military service, adopted in France in 1913, was a symbol of the competition of manpower. It increased the number of active divisions that would be ready for the initial battles.

The naval arms race between Great Britain and the German Reich was also based on the number of battleships in the fleets, although there was a qualitative race as well—guns against armour, mines against countermeasures, submarines against antisubmarine weapons, etc.

The permanent mobilization of science in the service of armaments since 1945 can be accounted for by the nature of the arms produced between 1939 and 1945 as well as by the tension that rose up between the victorious powers. The rivalry is based first of all, obviously, on the atomic bomb, or rather nuclear weapons; the atomic bomb was not conceived and built by arms engineers or technical specialists, but by genuine scientists, by theoreticians. It was Einstein and Szilard who signed the historic letter to President Roosevelt. Oppenheimer was the scientific director of the Los Alamos project that ended in the production of the bombs dropped on Hiroshima and Nagasaki; and theoreticians (Teller in the United States, Sakharov in the USSR) also played the most important role in the development of the H-bomb.

Up to now, I have used the expression 'qualitative arms race', but one could say simply that the creation of a scientific-technological complex devoted to military research brings about a constant improvement in armaments and the invention of new weapons systems.

Consider the simplest case: the first atomic bombs were heavy and cumbersome. It was obvious that engineers would work to improve the ratio of weight (or volume) to explosive power. The lowering of this ratio was not only the result of a sudden desire for elegance or economic efficiency; as the weight-explosive power ratio is reduced, the number of launchers, planes and missiles capable of carrying nuclear arms increases. In other words, miniaturization follows quite normally and logically from a rational calculation, military rationality as well as economic rationality.

In the development of nuclear arms over the last 35 years, two decisions were taken at the highest level. First was the decision to build the H-bomb. Here, one recalls the debate on this issue between Robert Oppenheimer and Edward Teller. The opposition of the former did not emanate mainly from a moral rejection of the bomb as such; but fissionable material was still scarce. Oppenheimer thought it preferable to concentrate first on the production of A-bombs. In the event, the Russians and the Americans exploded the first H-bombs within a few months of one another.

It appears that at the end of the 1940s and the beginning of the

1950s, political leaders were inclined to consider the problems of weapons almost exclusively from a technical point of view. To be sure, Truman and his advisers did not ignore the moral issues and they were immediately aware of the military revolution that the use of nuclear explosives had created. They tried by the Baruch plan to put this monstrous weapon out of the reach of sovereign states; but after the failure of this attempt, they threw themselves headlong into the production and improvement of nuclear arms; the technicians diversified both the weapons and the launch vehicles, for Truman feared that the Americans would be overtaken by the Russians.

The second decision taken at the highest level, this time in the 1960s, concerns what is called the MIRV technique, or multiple independently targeted re-entry vehicles in layman's language, the placing of several nuclear warheads in the missile nose cone. Here again, one could argue that the MIRV development follows from a quasieconomic rationale: it is not the missile that strikes but the warhead. One missile, therefore, can deliver warheads on three different objectives in the case of the Minuteman type III, or ten objectives in the case of Poseidon. But the decision to build MIRV missiles was not implemented without political and strategic controversy.

Between Truman's decision at the end of the 1940s and the MIRV decision by Kennedy and Johnson ten or fifteen years later, a number of political and technological changes had taken place. Instead of a unilateral deterrent—the United States forestalling Soviet aggression by the threat of a nuclear reply—there existed now a mutual deterrent: the Soviet Union was able to hit American soil with, first, bombers and then missiles. From the end of the 1950s—at the moment when Kennedy was affirming that there was a 'missile gap', which, however, did not exist—the Soviet Union had the means, although limited, to hit certain targets in the United States. The development of the missiles made the United States, in the short run at least, as vulnerable as the Soviet Union.

On the other hand, strategists, university men rather than soldiers, had conceived and elaborated the thesis of 'arms control', which, from this moment, accompanied the competition in armaments and led to continuous controversy.

How may we define 'arms control'? Officially, the theoreticians define it in the following way: a policy designed to *reduce the risks of war, the amount of destruction in case of war, and the cost of maintaining a military apparatus in peacetime.*

'Arms control' was the subject of studies that started mainly in the second half of the 1950s; atomic physicists participated actively in this strategic and political research. The original idea seems simple, almost obvious: the two superpowers perfect the same arms or the same weapons systems at the same time or a few years apart. Each one will have briefly, equal capacity to inflict unacceptable destruction on the other. Under these conditions, the arms race becomes absurd, since the race cannot be won by either side. Nor *should* one side win: for a victory would simply increase the danger of war.

But the necessity for disarmament does not follow from this reasoning; arms controls requires a limitation on nuclear armaments but it does not have disarmament as its primary objective. The theoreticians were thinking, above all, if not exclusively, of nuclear arms. They saw no chance of convincing the Americans or the Soviets to renounce all of their nuclear weapons. For, how could the elimination of these arms be verified? Moreover, the Americans would be the losers if these weapons were eliminated because of Soviet superiority in conventional arms. Once the hypothesis of an abolition of nuclear arms was put aside, what object could the arms control theorists have? They discovered a response: stability—this being defined as a situation in which neither of the duellists could be disarmed by the other, and, in consequence, neither would be tempted to have recourse to these arms because retaliation would be inevitable. So defined, stability seems to me the equivalent of, or the substitute for, the balance in conventional arms. Stability should be established at the lowest possible level, that is, the least costly. It is of little importance that one side is stronger than the other in terms of the number or quality of a particular category of weapons or launch vehicles; it is only necessary that retaliation by the victim of aggression be no less devastating than the aggression itself.

In this way, stability leads to the doctrine called MAD, or 'mutual assured destruction'. Effectively this means that it is sufficient for only a fraction of a country's nuclear arms to remain invulnerable so that it can survive an adversary's first strike, in order for the United States as well as the Soviet Union to possess the capacity to reply to a nuclear attack by destroying most of the enemy's cities. Certainly, if the cities of the United States or the Soviet Union were destroyed by a first strike, the United States, or the Soviet Union, would, so to speak, gain little by a vengeance attack. But in order for deterrence to work, it is important to maintain the certainty of a counter-attack.

The theorists of arms control and the commentators of news papers such as the *New York Times* have based their criticism of MIRV on the idea of stability. In effect, stability requires that a first strike be impossible; in simpler language, neither of the superpowers should have, or think it has, the means to take away from the other the instruments of retaliation. Any measure that puts in doubt the power to reply of the party attacked works against stability. Now the MIRV technology massively increases the number of nuclear warheads. And these numerous warheads make the other side fear a first strike that would disarm it.

At the present time, stability in strategic nuclear arms, supposing that this stability actually exists, is situated at a very high level, since both superpowers today possess about 7000 nuclear warheads. Even if the United States enjoys some superiority when bombers are included, this superiority in the number of warheads is offset by Soviet superiority in missile throw weight and in the explosive power of their warhead. The approximate number of warheads in strategic vehicles is 7000; the total number of warheads is nearer to 50000.

The present situation, the stability with 7000 warheads at the disposal of each of the superpowers, is clearly not the objective at which the arms controllers were aiming. For this stability is not maintained at the lowest cost; it would not reduce the extent of destruction in case of war. Nor is it even very stable since the heavy missiles of the Soviet SS18 could put the American Minuteman out of action.

Let us look again at the historical background. The missile race began at the beginning of the 1960s. The lead taken by the Soviets in the conquest of space with the first sputnik astounded public opinion in the United States and shook the confidence of Americans, wrongly as a matter of fact, in their technical-scientific superiority. The Americans replied with a massive, total effort, in their universities, laboratories and military programmes. The Kennedy team responded to the first Soviet strategic missiles with the programme of 1000 Minutemen, which, 20 years later, remain the land component of the triad of the American nuclear apparatus (550 of these 1000 Minutemen have been 'MIRVed'). The naval component comprised 41 nuclear submarines, each carrying sixteen missiles (first Polaris, then Poseidon).

The McNamara programme of 1961-62 was partly determined by an exaggerated estimate made by the intelligence services of the number of Soviet strategic missiles. It is possible, of course, that the intelligence chiefs really believed, at the end of the 1950s and beginning of the 1960s, that the Russians were going to build the greatest number of strategic missiles in the shortest possible time. In fact, it is known that the Soviets first produced medium and intermediate-range missiles aimed at western Europe. The massive effort concerning strategic missiles began a few years later and resulted, in 1971, in the signature of SALT 1, which accorded to the Soviets a superior number of missiles—1618 fixed missiles against 1046; 62 submarines against 41, and as regards the total number of missiles, 2358 against 1710. The Americans, however, had erred first by overestimating, then underestimating the Soviet potential. In particular, they overestimated the time necessary for the Soviets to catch up with them regarding the MIRV.

Why has arms control been inefficient? Are the thousands of warheads necessary? Is overkill an expression of madness? The commentators who search for a rationality in the arms race reply first of all that the more one of the two sides possesses missiles and warheads, the more the other must increase its stock of the means of retaliation. The warheads can be targeted on the enemy's launchers in the hope of at least dampening the reaction to a first strike. Nuclear stability, the object of arms control, requires that the weapons held for a second strike be invulnerable. Does not stability logically and out of prudence demand from both sides a high number of launchers or warheads?

When all of the planes of the Strategic Air Force were concentrated on three or four bases, they were clearly vulnerable to a first strike; when missiles replaced bombers, or were added to them, the invulnerability of the missiles in silos seemed, for a time, assured. But in the last few years, worry about the stability of the central balance has reappeared, and for a simple reason. Science has permitted progress to be made in one area which is of incalculable consequence—missile accuracy. The square error probable in relation to the target has fallen from some thousand metres to a few hundred metres. The Soviet SS20 missile can hit an objective thousands of kilometres away with a margin of error of a few hundred metres. The result is that the number of missiles necessary to destroy the silos of the other side is lowered. The number of missiles and warheads regains on importance it had lost when twenty warheads were required to put out of action a single missile in its silo.

These two consequences—the progressive multiplication of the number of warheads and their increased accuracy—which are perfect examples of the qualitative and quantitative arms race—did not stop, and were not even slowed down, by arms control. Neither of these developments could easily have been avoided. The first cut the cost of the weapon, the second flowed from the logic of technology. Be it a bullet, a shell or a bomb, the projectile must, in normal circumstances, hit its target; and engineers do their best to see that this is just what happens. They sought and found the means for long-distance accuracy just as they sought and found it for short and medium-range firing.

The effect of technical progress is, as usual, ambivalent. Such accuracy increases the vulnerability of the missiles (or any other land target) and, consequently, that of the implements of a second strike. On the other hand, the use of missiles or nuclear warheads does not inevitably mean the massacre of millions of innocent people. To be sure, however precise the firing is, nuclear warheads, launched in great numbers, would bring considerable collateral destruction. But this destruction would be reduced, all things being equal, according to the accuracy of the shot. To aim at a silo or an air base is less terrible than to target a city.

To this argument the theorists of arms control reply that any attempt to lessen the horror of the use of nuclear weapons is contrary to the political goal. In reality, the all-out promoters of arms control do not wish to see more accuracy in firing or an increase in the number of warheads, through MIRV. The number necessary for each of the two depends on the number of the other. The reduction of numbers on both sides is all that is conceivable. These two innovations, in effect, put into question the concept of mutual assured destruction. They risk undermining the certainty of a second strike because the number and accuracy of the nuclear warheads recreate a certain first-strike possibility.

The fact that the invulnerability of second-strike arms requires an increasing number of nuclear warheads makes clear the competitive mechanism of the arms race. But it does not make it rational from either the economic or military standpoint that stability be fixed at such a high level.

One can come up with another explanation, which is found often enough in writings on the subject: the military-industrial complex. To consider this viewpoint, one must suppose that a comparable complex exists on the Soviet side. In the United States funds for defence have at times increased and at other times decreased under the pressure of opinion and the authority of governments without the appearance, in a manner both visible and effective, of the representatives of this notorious complex. There is, nevertheless, some truth in the argument if it is put into different language. Naturally, officials of the different states of the United States put pressure on members of Congress in order that such and such a weapon be chosen and that an order be given to such and such a company. But the continuing fact of laboratory work, the permanent existence of research teams makes inevitable the improvement of existing armaments and the creation of new systems. Political leaders are not forced to order the production of all the prototypes, or all innovations, but they can hardly resist the temptation if they fear that the other side is building and implementing these innovations.

These two explanations of overkill—the mechanism of competition, and the level of continuing research—leave room, it seems to me, for a third: I am not convinced that civilian or military leaders are satisfied with the doctrine of mutual assured destruction backed by the arms control theorists and many scientists. They do not accept that any use of nuclear arms would inevitably entail an escalation to the extremes.

This thesis—that any crossing of the nuclear threshold would inevitably entail an escalation to the extremes—can be neither proved nor disproved. Physicists are no more knowledgeable on this subject than laymen or men of letters. The physicists' aim is to frighten political leaders and public opinion and deter them from using nuclear weapons, whether tactical or neutron bombs. Obviously, it is to be hoped that these arms will never be used; but is it advisable to tell the statesmen beforehand that escalation is unavoidable? Would it not be better to state the contrary? To reserve the possibility of stopping somewhere?

Moreover, this thesis that any crossing of the nuclear threshold leads to mutual assured destruction means that almost nothing is left for deterrence. If we follow the reasoning, the nuclear threat remains valid against a nuclear attack coming from another power but it does not work for anything else, and especially not for the allies of the United States facing non-nuclear aggression. Therefore the whole Atlantic Alliance must be reorganized on the principle of no first use. It is illogical, indeed, that the thesis of inevitable escalation should lead to the no-first-use doctrine.

The efforts against nuclear proliferation have not been totally in vain, however. They have made access to the atomic club more difficult. Of course, those nations determined to build nuclear arms will manage it whether or not they have signed the non-proliferation treaty, provided that they have the necessary technical and financial means: and the closing of the atomic club, although probably in accordance with the interests of mankind as a whole, gives rise to suspicion and resentment. It could be thought to hide ulterior motives so long as the competition between the superpowers goes on.

The arms control theoreticians can also take credit for several partial agreements, such as red line, the limited test ban (suppression of tests in the atmosphere, limiting the power of underground tests, forbidding the placement of arms in space or on the sea bed). But these agreements have not prevented the Russians and the Americans from working on the means to destroy one another's satellites and possibly to neutralize the whole of an enemy's satellite system.

The main agreements—SALT 1 and SALT 2, the last one signed but not ratified by the American Senate—involve the two superpowers exclusively. Both show the relations between the arms race and arms control. SALT 1 has been made more practicable by two technical achievements: verification through national means alone and the MIRV.

One of the obstacles to any military agreement is the refusal of such a secrecy-bound state at the Soviet Union to submit to on-site inspection. Satellites allow verification that is not perfect but is considered sufficient. The MIRV technology also helped SALT, but only temporarily. The United States agreed to a ceiling on the number of vehicles inferior to that of the Soviet Union because, thanks to MIRV, they possessed more nuclear warheads. But, as a whole, SALT 1 was a bad treaty because it authorized the modernization of missiles. Once the number of allowed launched vehicles had been set, SALT 1 was an incitement rather than an obstacle to progress toward the optimum use of each launcher. The Russians had accomplished less in the realm of miniaturization than the Americans, but they were able to make the most of the superiority of their missiles in throw weight. The largest missiles, once they had been 'MIRVed', could each carry megaton-power warheads (eight or ten, if needed) over thousands of kilometres. SALT 1 has not slowed down the arms race, rather it has shifted it in directions that are even more dangerous—that is, more warheads per missile, and warheads that are more powerful and more accurate.

SALT 2 is perhaps better in spite of the fact that it is not ratified. It is probably respected because neither one of the two superpowers is able or willing to do more than the treaty allows them to do.

The most spectacular success of arms control, the limiting to two sites of the ABM (antiballistic missile), is partly due to the scepticism over this system shown by most specialists. Is the absence of the ABM a contribution to the stability of nuclear balance? This stability is disputed more today than before. The large Soviet SS18 missiles can, on paper, destroy the American Minutemen. The Americans would be able to respond only with less accurate submarine-borne missiles.

Let us try to sum up the preceding exposition and draw some conclusions.

(1) If by arms race we mean the constant improvement of armaments, then the arms race, or in any case the qualitative arms race, is inevitable. This rivalry is determined and maintained not only by the nature of science itself, but also by the permanent mobilization of scientists and technicians in order to perfect current weapons or invent new ones. It would appear that several nations have entrusted laboratories with studying the possibilities of chemical or even biological warfare. It is perhaps possible that an epidemic has been caused somewhere in the Soviet Union by an experiment that either misfired or succeeded only too well.

(2) The most visible armaments race and the most discussed over the last 35 years has concerned nuclear arms and rockets. The achievements of engineers on both sides have been brilliant. The abundance of fissionable material has permitted the production of

thousands of warheads. The increase in the number of warheads resulted from economic-military logic and strategic considerations. The MIRV allows the number of warheads to be increased without augmenting the number of launch vehicles. But when the two superpowers mastered the MIRV technology, they did not reduce their stock of launchers. In any case, the number of one side's warheads forces the other to increase his own to insure the security of his second-strike capacity.

(3) Does it follow that it is impossible to reach an arms control agreement that would reduce the number of launchers or warheads and maintain the same stability, that is, the same mutually assured destruction? Such an agreement is certainly conceivable but its negotiation would be extremely difficult. The two SALT accords have imposed practically no restrictions on the Soviet and American programmes. How difficult it would be if the negotiators actually tried to agree on the destruction of several hundred launchers, or simply the freezing of the present situation; in any case, the reduction of the number of launchers will neither increase nor decrease the danger of war.

(4) Certain American commentators propose a radical change of direction: the refusal once and for all to participate in the competitive race. Since strategic nuclear arms should by themselves guarantee mutually assured destruction, the United States does not need 7000 warheads, and even less does it need to build new ICBM (intercontinental ballistic missile) systems. Bombers with cruise missiles and submarine-borne missiles constitute an assured second-strike capacity. Retaliation would conform to the definition of unacceptable devastation, the mutually assured destruction.

Let the Soviet Union ruin itself in this senseless contest; the United States should be satisfied with a minimum deterrent, as was said twenty years ago. Since the American goal is stability through mutually assured destruction, it can be obtained with less expense. Why should this not become the thesis of the United States?

Some will reply that it would be rash to disband the research teams, and close the laboratories. The other side might one day find a way to destabilize the mutually assured destruction, to employ new nuclear arms, to destroy the satellite system.

Moreover, could one of the superpowers resign itself to an apparent inferiority or a real weakness? Minimum deterrence implies the doctrine that any nuclear war must escalate to extremes and, therefore, that deterrence through the nuclear threat is good only against a nuclear attack.

The United States, without a common border with the Soviet Union, could, if need be, adopt this approach. But, in that case, what would be left of the Atlantic Alliance and the American deterrent? More generally, if any use of nuclear arms leads, accordingly, to the reciprocal destruction of the belligerents, one must draw the following inference: the western nations, Americans and Europeans, must profoundly modify their doctrine and prepare themselves for any eventuality that does not involve a recourse to nuclear arms. It follows logically that a race in conventional arms would result. If the West increases its armies, the Soviets would answer that the present balance is broken. To be sure, there has been a constant improvement in conventional arms as much as in the nuclear field. The supporters of the arms control thesis have dealt almost solely with nuclear arms. What should they say about conventional weapons?

(5) If the two superpowers agreed on a freeze of nuclear arms, one cannot be sure that they would reduce the total amount of their defence budgets. One of the aims of arms control—reduction of military spending—would not necessarily be achieved by stabilization, or even the cutting back of one part of the total defence budget. But the fact is that negotiations on conventional arms in central Europe have gone on for years with no result.

The partisans of arms control would reply that the most urgent task, the present duty, is to end the competition in nuclear weapons and to reduce the risk of nuclear war. All wars are terrible, but there is no common measure between a war waged with conventional arms and a nuclear war. States and nations can survive the former; but societies, perhaps nations themselves, would not survive the destruction of cities and populations.

This argument is undeniable. One can never repeat often enough that nuclear arms are essentially different from conventional weapons. But this argument is only convincing if one accepts the idea of an inevitable escalation. Would the use of a low-power nuclear arm be likely to bring about a total nuclear war? If one accepts the thesis of an irresistible escalation, the tactical nuclear arms, even neutron bombs, change nothing essential, they have no more meaning; the nuclear threshold must be maintained, it must be given the value of holy writ, governments of all states must be convinced that the first use of the nuclear bomb would be the beginning of the apocalypse.

(6) Based on this thesis, what does deterrence by nuclear threat mean? It means either bluff or suicide. Take the case of a small nuclear power, Great Britain or France. Let us suppose that enemy troops reach the French border and that the French government threatens to retaliate to an attack by conventional arms with the bombing of Soviet cities. If the enemy takes the threat seriously and does not cross the border, deterrence has worked. But if he calls the bluff, if he crosses the border, France would bring on itself a catastrophe, since the enemy, in theory the Soviet Union, would make France pay for the devastation it would have suffered, but at a much higher price. The French are not unaware of the part that bluff plays in deterrence theory. According to opinion polls, the French are for the most part in favour of the strategic nuclear force, but, for the most part too, they think that, should the enemy not be intimidated by this threat, the president would negotiate rather than press the button. A reasonable attitude, but up to what point would the enemy believe a threat in which the deterrent party itself does not really believe?

In the United States at the present time there is new thinking in favour of the no-first-use thesis. Public reaction in Europe against modernizing Euromissiles pushes certain commentators, personalities who once held high positions, to come out for no first use, arguing that the Europeans with a limited effort could re-establish an equilibrium with regard to the conventional forces of the two alliances, the Atlantic Alliance and the Warsaw Pact. To this it can be objected that in the past the apparent balance of forces did not prevent wars from taking place. By lowering the risk of using nuclear weapons, the danger of a war is increased, a war which, in spite of no first use, could become nuclear if it lasted long enough.

(7) What is the conclusion suggested by this analysis? I shall sum it up briefly: there are antinomies of strategy in the nuclear age just as there were antinomies in the *Critique of Pure Reason*, if I dare make this outrageous comparison. No one can be sure of himself and of his theory. By reducing one danger, another danger is inevitably increased.

To simplify, let us say that today there are two schools of thought; one is based on the idea that a nuclear war must be avoided at all costs and that the use of a single weapon may bring about total nuclear war. This school of thought pleads in favour of no first use and a European effort concentrated upon conventional rearmament. The other school thinks that between nuclear powers, when the stake is considerable, namely Europe itself, it is illusory to imagine that a possible recourse to nuclear weapons can be completely ruled out. This second school does not believe that any use of a nuclear arm, even a tactical one, means the end of the world.

Each school has strong points against the other; each is vulnerable to argument. After the thunderstroke of Hiroshima and Nagasaki, nearly 40 years ago, humanity was terrified by its own power. It could destroy itself with its own arms. Men of goodwill joined together to save humanity from a nuclear catastrophe. They decided—or events decided for them—to prevent war, in other words neutralize nuclear weapons as much as possible. Other men of goodwill, also eager to prevent war, decided to maintain the nuclear weapon threat in order at least to limit so-called conventional wars.

The second school still predominates with the theory of flexible response. The first school pleads in favour of no first use.

These controversies may be academic today, because a massive and frontal military attack coming from the east against western Europe seems unlikely. But the uncertainty persists. Can a nation, in the long run, base its security on a threat that is either bluff or suicide? Can it equate the use of nuclear arms with total nuclear warfare?

Can science or arms control overcome these uncertainties? I do not think so. The accuracy of missiles speaks in favour of the thesis that nuclear weapons can be used without an escalation to extremes. Arms control could perhaps lead to a freeze of the available arms of both sides (assuming that the campaign against proliferation is successful). Now, on this hypothesis, stability on the highest level demands a countervailing balance of the other levels.

Why should we be surprised by these failures? Science sometimes makes agreement easier and sometimes more difficult. What neither science nor arms control can achieve is to remove any danger of a war between nations whose ideologies and interests conflict and which submit neither to mediation nor to the international community. Enemy nations, because of the nature of nuclear weapons, seek to avoid destroying one another. Neither can accept a position of inferiority. Arms control wished to make the best of these two desires: a common desire to avoid total war and a desire on each side to be at least equal to the other. The results are unsatisfactory for a reason that is more related to human nature than to science: as Orwell wrote in *Animal Farm*, each one wants to be a little more equal than the other.

Many years ago, a colleague of mine defined my way of thinking in the following way: 'Professor Aron is demonstrating with convincing clarity that the world cannot be different from what it is'. I am afraid that this chapter does not contradict the definition. I present the arms race and the present nuclear balance as almost inevitable. I would accept the objection with some reservations. Firstly, it is doubtful if there was ever a chance of preventing the increasing sophistication of weapons development; but if there was, it was missed by SALT 1. Limiting the number of missiles without also taking steps to prevent their modernization has had negative consequences.

Secondly, I have reservations about the no-first-use doctrine which may come to be adopted by the Americans. Probably the majority of my audience feels some emotional sympathy with this doctrine. But there are two main objections to it. The first is that it implies an increase in the military budgets of the countries of the West, and no doubt the Soviet Union would increase its own military budget accordingly. Secondly, it would make possible conventional war in Europe. For Europeans, the campaign for the no-first-use doctrine could mean that, while American territory would be spared the horrors of war, Europe itself would become the theatre of war; and war, even when conventional, is terrible enough under modern conditions.

Thirdly, I should like to comment on a most interesting report from a group of physicists that I recently received. They argued that atomic weapons differ in essence from conventional weapons; that any use of nuclear weapons would in all probability lead to escalation; and that it would be relatively easy for the West to defend itself without nuclear weapons. There is, of course, nothing new in all this. We have heard it often before. But, the decisive point is the following one: if the West were losing a conventional war, it should not resort to nuclear weapons. A lost war, in the view of these physicists, is preferable to a nuclear war. Nations can survive defeat, but they would not survive a nuclear war. It is not exactly the famous or infamous slogan 'better red than dead', but it does make clear the meaning of the 'no-first-use' doctrine. It leads logically to the proposition—accept defeat rather than contemplate the use of nuclear weapons. What should our answer to this group of physicists be?

My final reservation is that any analysis of strategy brings us more or less to the conclusion, no way out (or *huis-clos*, as Sartre would have said). But despite that, we in Britain and in France *do* live without anxiety. We discuss the standard of living much more than we do the SS18. Are we blind or wise? For the last four decades there has been no war in Europe, despite the development of modern technology. Mankind is unable to eliminate or forget its capacity for self-destruction. When considering what should be done, we can only hope that Aristotle was right and that man is in reality a thinking being.

3 *The established dissenters: scientists and America's unwritten constitution*

DON K. PRICE

Professor Emeritus of Government and of Public Management, Harvard University

Spencer's reputation as a political seer—a prophet of perpetual progress and human perfectibility based on the advancement of science, and an advocate of the restriction of government to a minimum role in society—was in my undergraduate days perhaps at its lowest point.

Today, however, President Reagan's speeches sound like excerpts from Spencer's *Social Statics*, and Prime Minister Margaret Thatcher's occasionally have a faint flavour of the same source. But in the actual conduct of governmental policies, both our countries have adjusted to the difficult social problems of a technological era by exercising controls to an extent that would have horrified Spencer and his contemporaries. Yet while the two countries have responded in somewhat similar ways in the extent to which government deals with economic and social issues, they have done so by quite different political and administrative systems. As the British Ambassador to Washington, Sir Nicholas Henderson, remarked to the American press recently, 'You don't have a system of government. You have a maze of government . . . a whole maze of different corridors of power and influence,' which exist because the constitution was devised to avoid giving the executive ultimate power.[1]

Sir Nicholas summed up the difference between the American and British system in the conventional way, attributing it to the different constitutional systems in the two countries—a limited monarchy with a parliamentary system and an unwritten constitution, as against a federal republic with a written constitution and a separation of powers. It seems to me more realistic to look at the differences between these two systems of government not only from above, as a problem in sovereignty—the traditional perspective of the political theorist—but also from below, the seamy side of sovereignty, so to speak, with attention to the way institutions operate in practice, and as influenced by popular attitudes. From this perspective, it seems clear to me that some of the main differences come from the way in which science is related to politics in the two countries. If we consider the role of scientists in public life and the influence of science on political ideas, it may give us more insight into the development of political institutions both today and in the future.

I should emphasize that I do not think that the relationship between science and politics is the only, or even the main difference which explains (from the British point of view) American peculiarities. I have no idea how one could quantify the weight of its influence by comparison with the geographical extent of the country, its ethnic diversity, its strategic isolation, or the patterns of its economic development. All I would suggest is that scholarly observers have paid all these factors much more attention, and that the influence of science on political institutions has been comparatively neglected.

Toward freedom or tyranny?

The most fundamental issue with respect to the political influence of science is one that, fortunately, hardly need concern us if we concentrate solely on British and American institutions. At that basic level, Spencer was fully justified in worrying how the rigorous certainties of science could be reconciled with a system of political freedom. That worry today bothers the layman, if he takes seriously his George Orwell or Aldous Huxley, more than the scientist, since the scientist is more aware of the way in which the several specialized disciplines fail to combine into a single grand monistic system, a totalitarian synthesis that could provide the basis on which society could be governed scientifically. Three years ago Professor Holton's Spencer lecture discussed the competing themata of science that provide many internal degrees of freedom in scientific thought, and therefore a saving pluralism within the world of research.[2]

In the institutions and processes of government, there are less abstract and subtle issues that distinguish pluralistic or free systems from those committed to a totalitarian ideology, avowedly based on faith in science. A century ago, Spencer shared the view of some of his contemporaries on the continent of Europe—such as Auguste Comte and Karl Marx—that science was the mode of thought that would contribute to the progress of mankind, and that progress should continue onward to a state of social perfection. But then he broke sharply with their ideas, accusing Comte of following French ways of thinking—obviously a sin in English eyes—in wishing his ideal industrial state to be organized on a militant basis, in which government dominates the economy and 'the individual is owned by the State.'[3]

If we are concerned with the way in which we reconcile the approach of science to public issues with the systems of popular control and democratic accountability, we do not need—at least in Great Britain and the United States—to spend much time arguing against totalitarianism. In both countries, the major parties of the left and right alike are committed to democratic constitutional systems. Spencer's ideas about natural rights, and other elements of his dissenting heritage and scientific concerns, had been part of the systems of thought by which British and American radicalism on the whole split off from the European line of political evolution that led to Soviet communism. In neither country could the ideologues who wanted to base a dictatorial system on a set of scientific theories gain very extensive support.

Scientific freedom and administrative responsibility

On the contrary, scientists have been more generally associated with governmental processes that weaken the discipline of political parties, reduce discretionary authority in government, and (except when it has to do with their own research and teaching) favour more popular participation in both policy-making and administration. On these matters I speak with reasonable confidence only about the United States, but if I can deal sensibly with American experience, perhaps I can also raise questions about the way British experience is developing, and what problems it may cause, or benefits it may produce, for this country. The growing democratization of political and legislative procedures, and the reduction in the élite status of administrative generalists, may begin to give British scientists a role in public affairs rather more like their American counterparts. What may be the result?

The possible result might horrify not only those concerned with British traditions, but also those American political reformers who have sought to make our policies more coherent and our administration more efficient by imitating British institutions. If we consider the ways in which science is involved in American institutions for the development and determination of policy, it is clear that it is not congenial to the classic theory of Cabinet government. For example, Congress has insisted on setting up by statute in the Executive Office of the President a staff of scientific advisers, the Office of Science and Technology Policy, which President Nixon had managed to abolish and his successors were not apparently eager to have. Congress, among its 30000 staff members, has a special Office of Technology Assessment to help it understand the current impact of science on policy, but in addition it often, by statute, invites the National Academy of Sciences to give it advice on subjects with a high policy as well as scientific content. Most important of all, the career civil service, in its top levels, is heavily dominated not by administrative generalists but by officers who have come up through scientific, technical, and professional careers.

The influence of science, and the acceptance of science as a source of authority by politicians, are established in the American system, but not as a new feature, nor as the product of the fundamental scientific advances in the post-Einstein era. The origins of this way of thinking, and of the constitutional system that supports it, can be traced back to the first radical revolution of the modern world—the Puritan rebellion in seventeenth-century England. Historians of science have described the Puritan influence on the flowering of science after Bacon, and economists and sociologists have traced the way in which it encouraged the development of capitalism. But less attention has been paid to the ways in which that combination of religious and early scientific thought had an influence in shaping the fundamental American attitudes toward politics and government.

The Puritan Commonwealth was the first experiment that combined a written constitution (Cromwell's Instrument of Government), a republican executive, and a separation of powers, without an ecclesiastical establishment. All these heresies were suppressed after the Restoration, and almost forgotten after parliamentary leadership gained effective control in the United Kingdom during the eighteenth century. As long as the Stuart pretenders and the papacy seemed the main threats to good order in the kingdom, the Hanoverian settlement with its union of the landed gentry, the commercial interests, and the Anglican establishment were in firm control. But at the other political extreme from the supporters of the Stuarts were the remnants of those dissenting groups, including some of the leaders in the scientific institutions of the time, who not only continued to advocate the political principles of the old Puritan Commonwealth, but maintained communication with the leaders of the revolutionary movements in the American colonies.

There was nothing very new or original in the political ideas that were communicated; the clique of self-styled 'Honest Whigs', with whom Benjamin Franklin met in their fortnightly sessions at the London Tavern, were advocating ideas some of which were carried over from the Commonwealth of a century before: disestablishment of the church, a constitutional separation of powers to check the omnipotence of Parliament, and a federal system to protect the interests of Scotland and Ireland.[4]

Prominent among this group were leaders of the science of the period, for whom the new ideas of the Enlightenment combined comfortably with religious dissent—in particular Joseph Priestley and Richard Price, whom Edmund Burke denounced for their mixture of scientific rationalism and political radicalism. Priestley, a Unitarian and the discoverer of oxygen, in 1791 migrated to America after a mob supporting Church and King destroyed his home, and settled in Pennsylvania to take an active role in the American Philosophical Society.[5] It was Price, the statistician and Fellow of the Royal Society who had been invited to come to America as financial adviser to the rebellious colonies, who provoked Burke's outburst that 'the age of chivalry is gone. That of sophisters, economists, and calculators, has succeeded; and the glory of Europe is extinguished forever.'[6]

In the late eighteenth century, perhaps especially among dissenters and in America, the theologians and scientists were far more congenial in their thinking than they became in the nineteenth century. Cotton Mather, best remembered for his stupidity in dealing with the witchcraft mania of 1690, was a Fellow of the Royal Society, elected for his success with inoculation for smallpox and his experiments with plant genetics.[7] Jonathan Edwards, the leader of the Great Awakening, the New England Calvinist revival of the mid-eighteenth century, based some of his theological ideas on Newtonian science, and accepted the principles of material causation with no difficulty, holding that the 'laws of motion, and the course of nature' were so uniform that they might theoretically be predicted to the end of time by some 'very able mathematician.'[8]

Perhaps the best example of the harmony of religious, scientific, and political ideas in America at the end of the century was in the election of 1800, in which the two presidential candidates were the active heads of the country's two main scholarly and scientific societies, founded in imitation of the Royal Society. John Adams, who had succeeded Washington as President of the country, was President of the American Academy of Arts and Sciences, and Thomas Jefferson was President of the American Philosophical Society. The American Revolution had been motivated to a great extent by religious dissent; the rebellious New England colonies were outraged by the Quebec Act of 1774 which granted freedom of worship to Catholics in Canada, and the revenue measures of the Townshend and North governments seemed to be moves in the direction of imposing the Anglican establishment on the colonies. Jefferson, for all his deist scepticism, got some of his main political support from the extreme evangelistic churches, especially the Baptists who, after the disestablishment of the Anglicans, were eager to escape the leadership of the Congregationalists, who retained their status as established churches in Massachusetts and Connecticut until the 1820s.[9]

From that time to this, scientists have been allied with the political opposition to the creation of any central establishment, whether of career administrators or party leadership, that could help impose unified discipline on national policy. Their influence has contributed to the peculiarly American lack of coordination in government in three ways: (1) the fragmentation of the administrative system for the formulation and administration of policy; (2) the increase in both popular participation in government and in legalistic control over its detailed affairs; and (3) the moralistic resistance to compromises of the kind that enables a broad-based party to develop a programme without splitting into doctrinaire fragments.

These difficulties have had great compensations. They have freed science to innovate in public as well as private life in ways that a more disciplined system would have prevented, and to criticize political leadership with weight and authority. We may consider such benefits later, but let us first look at the ways in which science has contributed in America to the fragmentation of policy and of administrative institutions.

Fragmentation of the administrative system

Since the era of the French revolution, the most distinctive aspect of government in America has been the decentralization of its administrative institutions. In other countries, radical revolutions and the growth of democracy led to a centralization of administrative authority, with the breakdown of feudal institutions and the development of central bureaucracies. But in the United States, they decentralized the administrative structure. The main functions of government were in the thirteen original states (now 50 in number) and the cities, and within them the old constitutions and charters were amended to reduce central executive authority, subordinate it to legislative bodies, and provide for the popular election of not only the governors and mayors, but also of their administrative subordinates down to menial levels. As new functions of government came to require federal support, various expedients were found to pay for them out of federal taxes but to leave them to be administered outside the federal hierarchy, not by federal civil servants.

This process has continued up to the present, with a conspicuous effect: the more the government does and pays for, the less it moves toward socialism, in the sense of directly owning and managing the means of administration. We have not only failed to socialize the ownership of industry, but we have been steadily desocializing the administration of government.

Herbert Hoover once remarked sourly that while we had not socialized the ownership of business, we had—by the income tax—socialized its income. It is instructive to see what the money raised is actually used for. Less than a tenth of the federal budget now goes to pay for domestic programmes directly administered by federal civil servants. Rather more goes to pay for programmes contracted out to private institutions, and as much again for work done by state and local officials, and five times as much for funds paid out on statutory formulas, most notably social security payments to individuals.

Scientists played a part in the development of this fragmented system in several ways: (1) they had a major role and a highly beneficial one, in the initiation of new policies; (2) they helped devise systems of administration by which the federal government paid for new programmes but turned over their administration to other institutions; and (3) they dominated the development of the federal career civil service in patterns that resisted central coordination.

They worked in these directions because their scientific ethos committed them not to some unified system of basic science, but, in the best Baconian tradition, to an emphasis on practical applications in specialized fields.

In this respect, they were following the same drift of emphasis as the dissenting churches; lacking either an established church or a civil service establishment to control them, they pursued the practical and profitable courses of action. The Puritan churches, in their basic theology, were committed to an abstract faith, a confidence in the direct personal experience with Divine Grace that made unnecessary the intervention of any ecclesiastical establishment to guarantee salvation. But they also approved the influence of religion in its effect on society. It was only a purist like Cotton Mather who could complain in 1700 that the Puritans of Plymouth, Massachusetts, had degenerated from their original rigorous standards: 'religion brought forth prosperity, and the daughter destroy'd the mother'.[10] Others, over the generations, were less concerned with the abstract theology that separated the religious denominations, and more concerned with their combined practical influence in safeguarding morals and improving economic and social conditions. But above all, they were determined to maintain a system of free competition with their sister churches, to such an extent that there would no longer be any worry about the return of an ecclesiastical establishment.

It was by a somewhat similar process that scientists began in the early nineteenth century to influence the development of federal policies. If you wish to trace the origins of new public programmes in America, you will learn little from party platforms, but a great deal more from the proceedings of the principal scientific societies. The typical way new programmes were initiated in an era that believed in *laissez-faire* was for scientists and engineers to persuade

Congress to support some research programme, and later to convert it into an action programme, if possible outside the federal career service. This process was most effective when the scientists could lobby from a base in a private or state government institution, with the support of private or local funds.

For example, the American Philosophical Society persuaded President Thomas Jefferson to establish the first Coast Survey, the basis of future maritime development. The Steamboat Inspection Service was set up after Congress gave the Franklin Institute money for a scientific study of why steamboat boilers were blowing up, and the scientists went beyond their assignment and proposed the creation of the first federal regulatory programme. The growth of agricultural programmes came after the colleges for agricultural and mechanic arts in the several states were given federal grants to support research, and then converted the research projects into action programmes with more grants to the states and counties. Medical and public health programmes started with Rockefeller grants to county and municipal governments, and federal grants-in-aid were not long in coming.

It was this system of grants-in-aid which bribed the several state governments—or rather by which they bribed themselves—to abandon their pre-Civil War position of defending states' rights against the federal authorities, and to advocate an expansion of federal responsibilities on condition that the programmes be administered by the states. The high point of this movement came during Franklin Roosevelt's New Deal, in which the main outlines of his new policy had been developed by the economic and social sciences, especially in studies supported by President Hoover's Research Committee on Social Trends, financed of course not by the government but by a private foundation.

The next phase was the work of the scientists in developing a system in which private corporations took the place of state and local governments in carrying out federal programmes with federal funds—the new system of federalism by contract. This was, to my mind, the greatest invention of the scientists who worked on new weapons during the Second World War. The Office of Scientific Research and Development, led by the heads of the Carnegie Institution of Washington, the Massachusetts Institute of Technology (MIT), and Harvard University, took control of the most advanced programmes of weapons development away from the military services, to be dominated more by basic scientists than by practical engineers. Their entire programme was administered by universities and private institutions: it was the Massachusetts Institute of Technology that did the American work on radar, after the fundamental discoveries were made in Great Britain, and the Universities of Chicago and California, and the Union Carbide and other corporations that developed nuclear fission and the atomic bomb.

As a result of that experience, most of the new federal programmes since the Second World War have been set up on the administrative basis of contracts with private corporations. These have been not only those in which physical scientists and engineers took the initiative, such as the Atomic Energy Commission and the National Aeronautics and Space Agency, but those initiated by biological and social scientists—the National Institutes of Health, the programme of medical insurance, of housing and of urban redevelopment, and many others.

The government contracted out functions not only to established institutions, but to new private corporations created especially for the purpose, such as the Rand Corporation, and gave them functions involving research and the recommendation of new government policies, as well as the management of old ones. This is the corporate device that has also been used in Britain, where the so-called 'quangos'—quasi-non-governmental organizations—multiplied some years ago. But it may be significant that Prime Minister Margaret Thatcher has followed the advice of career civil servants in frowning on the 'quangos',[11] while conservative theory in the United States still finds them attractive. The difference, of course, lies in the different status of the civil service between the two countries.

It was less than a century ago that the United States created its first career civil service system, after reformers studied the Northcote-Trevelyan reforms in Whitehall. But the anti-establishment prejudices in Congress led it to enact a quite different system; it prohibited any civil service regulation that would forbid any citizen at any age to enter the career service and rise to its top levels, and thus made it impossible to imitate the very features of the British system that made for an influential higher civil service. The United States could not legally have any requirement for a permanent career system with entrance at an early age, or a separate category of higher-rank officers, or any educational requirements except in scientific or professional categories. As a result, the scientists and professionals, easily superior in competition with officers brought in by political patronage, began to move up into the highest ranks. Today, about two-thirds of the jobs in the top three grades of the service are occupied by those with scientific, professional, or specialized education and experience, and the movement back and forth between government and private employment is possible at all ages.[12] Much as Calvinist theology by abolishing episcopacy three centuries ago put the careers of the Puritan clergy at the mercy of local congregations, so the American theory of the civil service puts each position at the mercy of market forces.

Perhaps most important of all in preventing the development of a civil service that could bring more coherence and continuity into American policies, there are two general political prejudices that colour the attitudes of Congress toward the civil service—and a British audience may need be reminded that in the United States all the conditions of organization, recruitment, and employment of the civil service are determined not by executive action, but by Congress. The first is the myth that the civil service is supposed to administer policy determined by law, but not to play an active role in formulating or advocating it. In practice, of course, this idea restricts the civil servant mainly with respect to the more important interdepartmental issues, but even at lower levels of policy formulation it puts him at a disadvantage in competition with advisers from outside the government. The second is that no civil servant is permitted to occupy a position in the hierarchy of any department that puts him over all other civil servants in that department; no department, much less the government as a whole, is permitted to have a head of the established career service.

In recent years there have been efforts to move in a limited way toward a system in which there is something more like a unified higher civil service, with more promotions across departmental lines, to work toward more continuity and coherence in policy. Those efforts were delayed and restricted by the general resistance of the scientists and professionals who dominated the higher service. Here I need not spell out in detail the contrast between the American system, or lack of system, and the British civil service. But it is interesting to an American observer that in the United Kingdom there are those who would like to loosen up the control of the career generalists in the service, and bring more scientists and professionals into the upper ranks, as suggested some years ago by the report of the Fulton Committee.

American scientists, with all of their advantages over the generalists while serving within the career service, have still more opportunity to exercise influence over policy from positions off the federal payroll. For the system of federalism by contract, and the programmes of federal grants for the support of research, have created an alternative establishment outside the hierarchy—an establishment of dissenters. It is the scientists from private institutions who are most readily accepted as advisers on high policy matters, either as members of part-time committees or on short-term leaves of absence from their home universities. This dissenting establishment is maintained by systems of federal financial support that are carefully calculated to insulate it against governmental control, in spite of the Erastian efforts of budget examiners and inspectors general.

It was clear to those who developed the system of support for the National Science Foundation and the National Institutes of Health a generation ago that in America nothing like the University Grants Committee of Great Britain could be relied on to insulate private universities from political meddling. The experience with the biggest programme of research support before the Second World War—that of the Department of Agriculture—warned against a programme that, on egalitarian and practical political grounds, would soon be controlled by congressional committees, through statutory quotas based on geographical or population formulas, and then distorted by local pressures, without much regard for basic science. The result was the system of grants awarded competitively for the support of individual research projects, with the competition judged by committees controlled by the scientists from universities, selected for their leadership in their scientific specialties. As long as the actual selection of specific projects could be controlled by scientists brought in from outside the government, it would be safe—for purposes of asking for money from Congress—to group those projects into programmes, labelling each one as designed to contribute in the long run to some particular practical use in technology or medicine. The scientist did not need to know what applied purpose was the justification for his grant, and the politician who supported its purpose did not need to understand whether there was really any connection between it and the basic science that it was supposed to justify.

Populism and legalism

The second characteristic American attitude to the role of science in politics is a combination of populism and legalism. This, too, had early roots; it was cultivated by religious dissenters to prevent the growth of an ecclesiastical establishment, and is used today to give the dissenting establishment of science a favoured channel of influence on policy issues.

The key to this influence lies in the institutional arrangement that is defined by the misleading metaphor of 'separation of powers'. The United States has an arrangement that is nothing like a neat separation; by contrast with the British government, with its rather clear separation of the processes of legislation from the control of the executive and administrative functions, the constitution leaves the powers entangled in a complete muddle. Congress has, in both law and practice, control over the system of administration—its organization, finance, and personnel. The strongest presidents are those who, like Roosevelt and Reagan, take little interest in management and generally leave it to congressional committees, while concerning themselves mainly with leadership in legislation.

This system is rooted in popular belief in two apparently conflicting principles—direct democracy and the rule of law. The origin of this paradox was in the theology of the dissenters. They believed in the salvation of the individual by divine grace rather than by obedience to the law or the authority of priests and bishops. And they believed that every believer should be encouraged to read and interpret the scriptures for himself. But then, to escape the danger of antinomianism and anarchy, and remembering the excesses of the Levellers in England and of the Anabaptists of Munster, they subjected themselves to rigorous rules of conduct. The extreme democracy of the New England town meeting worked against any established authority, and as the towns had to join in larger units of government to handle wider issues, they insisted on settling those issues by legal definition and judicial interpretation, rather than by accepting administrative authority. The way to let the will of the people control a complex issue was to spell out the answer in detailed legislation, and then let the lawyers and judges decide what it meant.

The early Puritan theologians and the later Jeffersonian sceptics were generally inclined to think that science would be a liberating political force, weakening traditional authority and strengthening democracy. This idea encouraged the modern hope that the general public should be educated in science, so as to avoid the danger of domination by an élite. At its absurd extreme, this attitude suggested that some day every major issue could be presented to the general public for decision by an instantaneous electronic referendum. In its more feasible form it led to the creation of numerous associations or organizations, including both lay and scientific members, for educating the public and Congress on crucial issues in which science was significantly involved. Television commentators, philanthropic foundations, and government agencies combine to encourage or support such organizations as the Federation of American Scientists, the Council for a Liveable World, the Union of Concerned Scientists, the Society for Social Responsibility in Science, and hundreds of others concerned with the prevention of nuclear war, the protection of the environment, or the control of genetic engineering.

These many dissenting societies compete as advisers with the established church of American science, the National Academy of Sciences and its subsidiaries—the National Research Council, the Institute of Medicine, and the National Academy of Engineering. These élite organizations, together with the various research institutes or 'think-tanks' supported by government funds, have special access to the channels of political advice, but by no means a monopoly on it. They are on call to help—sometimes to concur and sometimes to criticize—the top advisory authorities to the President and Congress. These are the Office of Science and Technology Policy in the Executive Office, and the Office of Technology Assessment in the Congress, both staffed in their top levels by short-term or part-time recruits from private research institutions.

What gives the scientists their greatest influence, however, is not their relationship with the President and Congress as a whole, but with the subordinate and uncoordinated parts of the executive and the legislature. The 500 subcommittees of Congress, each with authority to select its own staff, and without the modest degree of party discipline that was formerly supported by the seniority system, exercise 'oversight' over the executive bureaux (in any other government it would be called administrative control) and have direct access to the policy advice of both subordinate civil servants and outside scientific advisers. Their power today depends on the internal congressional rules of procedure as well as on constitutional law.

In the Government of the United Kingdom, the unification of power depends on the procedural rule of the House of Commons, in essential effect since 1707, that the House will consider no grant of funds that has not been requested by the Government. In Congress, by contrast, the power of subcommittees is reinforced at two levels. At the top, Congress is free not only to appropriate funds that the President has not asked for, and does not want, but to force him to spend them. If Congress wishes to have one of its subcommittees instruct an executive bureau on the way in which any expenditure is to be administered, that subcommittee is seen by the executive bureau as in control of the process of granting or withholding the necessary legal powers or financial appropriation for the next year. And if Congress is willing to let a legislative subcommittee assume absolute control over the initiation of new individual projects—such as a public works or flood control project—it may by an internal procedural rule require that the subcommittee's approval will be necessary before the appropriations committee may even consider a grant of the necessary funds.[13]

The member of a congressional subcommittee is likely to be more concerned with the interests of his constituents than with national policy. He must live in his constituency; the eighteenth-century Puritans held to the medieval theory of representation by which the representative's obligation ran to his district, and not to the King or to his party leaders. Today, if he wishes to impose the will of his subcommittee on a recalcitrant executive bureau, he is likely to seek support by getting scientists as consultants to criticize executive policy, and then to include in statutes, in precise scientific terminology, the standards which the judiciary enforce on administrative discretion. As a result, presidential leadership in policy, followed up by control of the administrative measures designed to carry out that policy, is extremely difficult.

The historic principle of British monarchy is that the King can do no wrong; the King or Queen may be as sinful as any subject, but for any public act the Ministers of the Crown are responsible, and must have monarchical power to carry out decisions as long as they are supported in office. The equivalent American principle is that the President can do no right; he may have tremendous support and esteem, but even if he can get Congress to enact a major legislative policy, he cannot assume that he will get the legislative support he needs for the administrative means to carry out the policy, much less for the adjustment of other policies and programmes that are related to it.

Moralistic resistance to compromise

Such coordination is hampered in the United States by the ways in which administrative institutions are fragmented, and executive authority restricted, by popular attitudes in which scientists play an influential role. Their effects are intensified by the moralistic attitude that gives little weight to the necessity for compromise when related programmes of government involve conflicting values.

In this respect the typical scientists, whose career has been confined to a specialized discipline and to research institutions, is likely to have an outlook on politics and administration that is in some respects—for all their difference in beliefs—rather like that of the dissenting clergy of a century ago. Each type of career was the main calling of the intellectual whose self-esteem depended on a sense of superiority to the moneygrubber in business or the compromiser in politics.

For the clergy in the American evangelistic denominations, this sense of self-esteem depended also on a defence of the fine points of doctrine that distinguished his particular sect from its competitors. Similarly, for the American scientist who takes an interest in some aspect of public policy, it is hard to avoid the delusion that his specialized key to knowledge should be the guide to decisions. Since most policy issues are complex, and involve issues that turn on several scientific disciplines as well as on a whole range of unscientific value judgements, the scientist is tempted by his loyalty to his discipline and his colleagues to an uncompromising view of the issues.

For example, in environmental regulation, the United States and Canada have a difficult problem as they discover that the pollution from midwestern smokestacks produces acid rain that kills fish and wildlife and is harmful to health in other ways in the north-east. The physicist or chemist who measures the pollution, the engineer who is called on to change industrial practice or smokestack design, the biologist who has to calculate ways to minimize the damage, the economist who must study the costs of prevention in terms of industrial losses and governmental deficits and perhaps increased unemployment—each may think he has the answer to the problem. But of course the problem is not exclusively scientific, in spite of the sophistication of the modern policy sciences. Issues of fundamental values—freedom and distributive justice—come into play, and so do issues of the distribution of political power.

In this kind of complex situation, the scientist—even more than the evangelist—is tempted in two ways to take a moralistic stand against compromise, and in an additional way is exploited by politicians.

His temptation comes, first, from his training in precision of thought, which makes him uncomfortable with any effort to solve a problem in non-quantitative and unsystematic terms, which are often the stuff of politics. His discomfort is most pronounced in dealing with the highest level of policy. For the reductionist approach of science makes its competence most apparent at the lower levels in the administrative hierarchy. That is a tautological statement: if a problem is one that can be answered conclusively by an exact science, it need not be referred upward in the hierarchy for decision by top political authority.

His temptation comes, second, from something like a sense of collective guilt. For two or three centuries the community of science had held to the millenial faith that science would guarantee perpetual progress. Now, after two world wars and the invention of weapons that carry the danger of destroying all civilization, and developments in biotechnology that seem to open up the possibilities of unpleasant controls over human freedom, scientists spend more time worrying about their possible responsibility for disaster than congratulating themselves about progress. And their conscience is not eased by the reflections of some of their more philosophical colleagues who no longer assure laymen that science is merely an extension of common sense that can serve as the basis of decisions in a liberal democratic system.

These pressures are made far more troublesome by the way in which American political leaders exploit scientific authority in support of their rival positions. Cynical politicians in the early years of the century took advantage of the evangelists' campaign in favour of the prohibition of alcohol; those members of Congress who voted dry and drank wet found it comfortable to call on the ministry in support of their public positions. Today each of the rival leaders in many a congressional controversy (energy policy is a good example) cite as impartial scientific authorities the dogmatic specialists who support them.

The American scientist who takes an interest in policy issues is tempted to take an uncompromising position by the reductionist approach of his training and the incentive system of his career. If he is a short-term or part-time adviser to goverment, he loses nothing by standing by the creed of his disciplinary specialty, and seeking the perfect solution to a messy political problem. This outlook was the one that Edmund Burke denounced, in his *Reflections on the Revolution in France*, saying that the essence of politics was 'in compromises between good and evil, and sometimes between evil and evil', calculating on a 'computing principle'—but 'not metaphysically or mathematically, true moral denominations'.[14]

The computing principle and constitutional change

Something like Burke's position, of course, was the basis of the prevalent English approach to government, and especially to the nature of its career civil service. The Anglican established church itself depended more on the humanistic and generalist education of Oxford and Cambridge, and less on rigorous training in theology, than its counterparts in either Scottish Presbyterianism or the Roman Catholic Church.[15] Following its example, the civil service establishment recruited men with education in humane disciplines, and promoted those who learned not to be stubborn in defence of specialized policies, but to help political leaders compromise their policies with each other, in order to deal with the long-term problems to be faced by both parties. As the association of higher civil servants told a Royal Commission more than 50 years ago, it is their duty to weigh mere 'Parliamentary convenience of today' against 'the steady application of long and wide views . . . which will endure or emerge long after the period of office of the Government . . . under whose authority it is taken'.[16]

Americans intent on political and administrative reform have long looked enviously at the cohesion and continuity of policy which is produced by such a generalist civil service and disciplined cabinet system. Such cohesive institutions could not develop in a system that was so deeply fragmented by the dissenting establishment of science, following the early influence of religious dissenters. Such fragmentation could not be repaired by the frequent proposal to amend the American Constitution by making the tenure of the President dependent on majority support in Congress, which would only give the committees and staffs of Congress even greater leverage in breaking up the unity of the Executive.

It may well be possible, however, to continue the process of step-by-step reform that, by comparison with several decades ago, has strengthened the central administrative institutions and the political leadership of the President. To continue such reforms requires Americans to recognize, as does the parliamentary tradition, that the scientific approach cannot calculate completely the comparative merits of various approaches to public policy issues. For any such issue combines elements that can be measured and calculated, and those that cannot—such as questions of basic legal and civil rights, and of changes in policy to take into account the competing interests of various regions and classes of society. They also involve the even less quantifiable questions of political power and partisan advantage. In putting together these elements of any issue with its quantifiable elements, it is essential in a democratic system to let general political judgement prevail over the calculations of the scientist.

America would do well, as it continuously revises in practical ways the workings of its unwritten constitution, to move further toward the disciplined responsibility of the parliamentary system. But, much as I would like to see the United States move in that direction, a move that would certainly require some reduction in the independence of action of scientists within the government service, I would still wish to hold any such move within limits, and to retain some of the advantages that go with the looseness in the system. For even though scientific knowledge cannot fully answer any complex policy question, it can do much to develop the answer.

The contribution of science to policy becomes more and more necessary in an age of rapid technological development, when economic issues turn on the new contributions of engineers, and social issues on new developments in biology and medicine. A ruling party and its career subordinates will often be tempted to put consistency ahead of innovation, especially in order to defend themselves against the demands of those regions and classes outside the dominant political interests of the country. This temptation may sometimes be restrained if independent scientific authorities are given a chance to criticize existing policies, and to suggest new ones that might be given public consideration by independent legislative committees.

Such an independent approach, even if it is a threat to the disciplined responsibility of a cabinet system, may have several virtues. It may let the scientists who are more aware of possible innovations bring up ideas for public consideration that the cautious generalist civil servant, or the harried party whip, may prefer to suppress. It may keep issues open on which the claims of competing value systems—those of diverse regional or class interests, or those involving civil or legal rights—may be in conflict with the utilitarian calculations of the majority. And if any type of critic may be permitted to disturb settled majority policy, the scientific critic may be the safest, since scientists, while by no means always disinterested, are generally unsuited by temperament to become rival political leaders.

Such an open (and sometimes irresponsible) approach, even though in America we overdo it considerably, seems to me one that it is essential to continue within reasonable limits in a federal republic. Is the United Kingdom an entirely different matter?

For an American who tries to follow British political developments from a distance, it is fascinating to see that the growing democratization in British society, and the increasing importance of science and technology, are leading to experiments that are something like slight steps in the direction of America practice. The development of specialized committees in the House of Commons, the continuing efforts to increase the numbers of scientists in top administrative positions, and the experiments with the use of economists and systems analysts from outside the civil service as advisers to the Prime Minister and cabinet, would all have astounded me as a young student here in the early 1930s.

Within the limits set by the different systems of sovereignty each of our two countries seems to be adjusting its practical arrangements somewhat in the direction of the other's. Each has found congenial this pragmatic type of compromise on various crucial issues on which theorists are in bitter disagreement. Neither country, for example, in adjusting its economic system to modern reality, has felt obliged to choose between unrestricted capitalism, such as I suppose Herbert Spencer would have approved, and pure socialism, along old-fashioned Marxist lines. If practical compromises may be made between those two ideal systems, perhaps similar compromises may be needed between the ideal of a unified and disciplined executive control of the legislature's proceedings, and one that puts all aspects of administration at the mercy of a fragmented collection of legislative committees. Now that government has enlarged the scope of its responsibilities to include management of the economy, it is harder to defend the theory that everything that it does should be the responsibility of a single collective will, guided by a comprehensive party doctrine. There is too much variety in the scientific and technological developments that alter old procedural habits, and open up new solutions for old political problems, for government itself to remain so completely unified in the way it maintains democratic control of executive power.

Is it possible to maintain the essential responsibility of the Cabinet to the House of Commons and still loosen up the dominance of the generalist in the higher civil service, and of the Cabinet in parliamentary proceedings? On this issue, if I were a British subject, I would be inclined to caution, since legislative encroachment on the executive, as the experience of some republics that have adopted parliamentary regimes suggests, may be harder to resist where the executive does not have the protection of a fixed term of office.

If there are compromises to be made on this issue, they may well be calculated on Burke's computing principle, which was prepared to admit that the essence of politics sometimes required compromise even among competing evils. I suspect he would have considered evil both the growth of democracy, which has destroyed much of the political deference that Bagehot considered useful for a Cabinet system, and the growing influence of science in policy, which challenges the administrative dominance of the generalist élite. But he did once point out (quoting Bolingbroke) that 'you can better ingraft any description of republic on a monarchy, than anything of monarchy upon the republican forms'.[17]

On this principle, Great Britain may have more leeway than the United States in reconciling central authority to the use of scientific expertise. It is not always by rigid defence of authority that its essence can best be defended. The idea of party government with a loyal opposition seemed to the old regimes of Europe a shocking threat to legitimate authority. In the end, with the growth of democracy, it surely made authority more acceptable.

The challenge of the new science and technology to old party ideologies, and to the dominance of old administrative élites, may be upsetting to the traditional temperaments of Westminster and Whitehall. In the long run, it may well add to the authority of parliamentary leadership and the career civil service alike. But if it is to do so, the scientists themselves will need to accept more clearly the limitations on their range of expertise. Only if some of them become generalists themselves will they earn the right to move into roles of leadership in both politics and administration.

Acknowledgements

In addition to the specific sources cited in the notes that follow, I should acknowledge the help I received in interpreting recent developments in American government from the study of the Presidency conducted in 1980 by the National Academy of Public Administration. That general study was supported by several specialized studies, and I relied especially on those by Hugh Heclo (who served as director of research for the NAPA study), Calvin Mackenzie, and Frederick C. Mosher.

Notes

1. *New York Times* 21 April 1982, p. A20.
2. Gerald Holton, 'Thematic presuppositions and the direction of scientific advances', in *Scientific explanation*, p. 4, Clarendon Press, Oxford (1981), undertakes to explain 'why is science not one great totalitarian regime, taking everyone relentlessly to the same inevitable goal?'
3. H. Spencer, *Principles of sociology* (1876), quoted by J. D. Y. Peel (editor) *Herbert Spencer and social evolution*, pp. 190-9, University of Chicago Press (1972).
4. Caroline Robbins, *The eighteenth-century commonwealthman* p. 336. Harvard University Press (1961).
5. Henry F. May, *The enlightenment in America* pp. 219-22. Oxford University Press, New York (1976).
6. Richard Price, *Observations on the nature of civil liberty, the principles of government, and the justice and policy of the war with America.* London (1776). Reprinted in facsimile edition by Dunderave, Ltd. (1976). See especially pp. 8-9, 12-13, 15, 99-100.

See also Henri Laboucheix, *Richard Price* (Montreal, Paris, Brussels: Didier, undated, 1970?), especially pp. 29-44. Price as an early social scientist was an early victim to the temptations of his successors. His statistics were sometimes biased by his political prejudices, and by his emphasis on scientific method he sometimes ignored the essential policy problem. Turgot, who had lost his position as controller of finance for the King of France, wrote to Price that his calculations failed to take into account the complexity of the real world, and that 'cette science approfondie serait plus intéressante pour les philosophes qu'importante pour les politiques . . .' Laboucheix, p. 33, n. 30.

See also *Reflections on the Revolution in France*, 1790, and on the proceedings in certain societies in London relative to that event. In *The works of the Right Honourable Edmund Burke*, Vol. IV, p. 83. The World's Classics, Oxford University Press (1934). Price referred to this attack in 1790 in the preface to the fourth edition of *A discourse on the love of our country.*

7. A general comparison of the status of science in America in the eighteenth and nineteenth centuries is given by I. Bernard Cohen in *Science and American society in the first century of the Republic.* (Alpheus W. Smith Lecture at the Graduate School, Ohio State University, Columbus, Ohio, 1961).

8. Alan Heimert, *Religion and the American mind* pp. 74-5. Harvard University Press (1966).

9. J. R. Pole, *The pursuit of equality in American society*, pp. 79-83. University of California Press (1978).

10. Cotton Mather, *Magnalia Christi Americana* (ed. Kenneth B. Murdock) p. 143. Harvard University Press (1977).

11. Robert Presthus, 'Mrs Thatcher stalks the quango: a note on patronage and justice in Britain.' *Public Administration Review,* **41**, (3) (May-June 1981), pp. 312-17.

12. Frederick C. Mosher, *Democracy and the public service*, Chapter III. Oxford University Press, New York. (1968).

13. *Hearings before the subcommittee on the separation of powers of the committee on the judiciary*, US Senate, 90th Congress, US Government Printing Office (1967). For a discussion of the relation between congressional intervention and administrative abuses, see Louis Fisher, Congress and the President in the administrative process: the uneasy alliance; in Hugh Heclo and Lester B. Salamon (eds.), *The illusion of presidential government,* pp. 35-40. Westview Press, Boulder, Col. (1981).

14. Edmund Burke, *Reflections on the Revolution in France*, pp. 67-8.

15. E. R. Norman, *Church and society in England 1770-1970* Clarendon Press, Oxford (1976), especially pp. 8-12, discusses the extent to which the Church of England clergy owed their social ideas to an ordinary university education rather than to any specialized theological study.

16. *Royal Commission on the Civil Service 1929-31* Report. Cmd 3909 (HMSO, 1939). Appendix VIII, pp. 5-7.

17. Edmund Burke, *Reflections on the Revolution in France*, p. 138.

4 *An old and intimate relationship*

MARGARET GOWING

Professor of the History of Science, University of Oxford

In December 1980 the retiring President of the Royal Society, Lord Todd, devoted his farewell address to relations between science and government and the major role the Royal Society itself should play therein. In fulfilling this role he said that the Society's objects were threefold:

1. To protect and encourage science in all its aspects pure or applied.
2. To offer to government an independent source of advice and help in the creation and operation of instruments through which science and technology might be brought to bear upon the formulation of national policy.
3. To develop international scientific relations, upholding the principle that scientists should be free to interchange their findings and to collaborate in the search for knowledge without let or hindrance.

To realize these objectives, said Lord Todd, the Society must not only continue to maintain its independence and its high standards but must—and here is the punchline for the purpose of this chapter —'*avoid involvement in politics*'. Lord Todd probably meant by politics, party politics and indeed he said that the Society's independence had been affected by the political commitment of Lord Blackett, the physicist, to the Labour Government during his Presidency (from 1965 to 1970).

This statement of objectives contains the most enormous claims for science and the scientists. Do we really want to encourage science in *all* its aspects? Professor Aron and Sir Alan Cottrell in their chapters in this book have spoken of the role of science and scientists in the terrible technological escalation of nuclear weapons. Do scientists in practice really demand freedom to interchange their findings? How many thousands of scientists voluntarily sign the Official Secrets Act as a necessary condition of their employment? Could, or would, the Royal Society freely pursue scientific collaboration with a country with which all official diplomatic relations were broken?

Beneath these claims on behalf of science, there flows, I believe, an old and constant theme about the relationship of science and politics: to put it crudely, that science is pure and politics is dirty. This after all was also the view of that great and highly political scientist, J. D. Bernal. The scientist might, he wrote in 1939,

> become a politician but he will never become a party politician [because] he sees the social, economic and political situation as a problem to which a solution must first be found and then applied not as a battleground of personalities, careers and vested interests. . . . Only when the parties can get together on a broad programme of social justice, civil liberty and peace can the full help of the scientist be expected.[1]

It is fair to add that after the Second World War, when Bernal himself was very deeply and successfully involved in military planning, he saw that the responsibilities of the scientists and politicians now intermingled so inextricably that the two groups must, he argued, understand each other.

What Bernal now realized was that Lord Todd's dichotomy between, on the one hand, government and science policy in all its different meanings—by inference a noble and necessary business—and on the other hand politics—by inference dirty and devious—does not exist. The words policy and politics after all have a common linguistic root in the Greek word for citizen and citizenship. Politics is the process by which groups of people, often with widely differing objectives and widely differing views of how to meet any single objective, come together to form governments, and governments are essential to identify and carry on the common purpose of a group—whether at the level of national or local citizenry, in academic societies, trade unions or whatever. Politics is about the relationship of people, about authority and power in society, about opinions and principles and choice. It is the very essence of policy formation, of the art and science of government.

So science policy means science politics. The lectures in this present series are testimony to the involvement of science with the politics of choosing ends and means. There are no doubt some policy questions where reputable scientists are unanimous in their analysis of scientific facts and their assessment of ends and means, but fewer than is commonly supposed.

Even if unanimity usually, or often, prevailed, science policy would still be enmeshed in politics because it requires money—lots of it. Money has been at the core of political debate and struggle in every community since the dawn of history for it is the key to the management and division of resources within society. Now that governments are by far the largest source of cash for science, most scientists, whether they are supported by government establishments, research councils or universities are involved in politics—that is in asking taxpayers to transfer resources from their pockets to scientific purposes. How horrified Herbert Spencer, whom this book of lectures commemorates, would be by this expenditure on science. It was the keystone of his political philosophy that governments should spend as little as possible on everything except law and order and defence. He could not have foreseen the strange circumstance that after the Second World War, in the words of Isadore Rabi, the physicist, 'The Navy saved the bacon of American science'. The Navy poured out money not just for work of clear relevance to warships but for much of American pure, basic research.

It is not, that is, only applied science which is mixed up in politics. To take another example of politics and pure science: for many years by far the biggest item in the British Science Research Council budget was for CERN, the European high energy physics centre at Geneva. This centre was not created from a strong conviction of the importance of such physics but because it was believed—long before the Common Market—that there should be a scientific contribution to the idea of European unity, that German and Italian scientists must be brought back into the European fold. A branch of physics must be chosen with no practical use whatever, but it was called nuclear research in order to open the pursestrings of government. Hence the curious fact that the letters R and N in the name CERN stand for *récherche nucléaire* when no actual nuclear research is done there. Other claimants on the science research budget are only too conscious of the compelling political as well as scientific momentum behind CERN's claim for cash.

Today the intimacy between science and politics should be strong and apparent. But, some say, this is a very recent relationship dating only from the Second World War, and contrasting strongly with the innocent, academic paradise of prewar science.[2] This belief in a paradise that has been lost and might be regained is, however, a fantasy at total variance with history, which shows on the contrary that the relationship is not only intimate but very old. I hasten to add that in demonstrating this relationship I do not suggest that it is the only, or necessarily the most important, thread in the history of science.

The rest of my chapter is in two parts. First, I shall indicate—inevitably very impressionistically—aspects of this intimacy from the seventeenth century to the outbreak of the Second World War. Secondly I shall show how the Second World War brought a new relationship between science and politics when, in the development of the first atomic bombs, scientists became very directly involved in the balance of power and the life and death of nations.

The historical relationship between science and politics has often had religion as a third point of a triangular relationship. One obvious theme might be the role of science in the secularization of states, whether in the seventeenth century—when the ideas of Kepler and Galileo helped to change the political as well as the theological foundations of a system of European government based on the papacy—or in the eighteenth century, when the American and French revolutions created secular states, or in the nineteenth century when Darwin's theories seemed the culmination of a progression of new ideas which undermined the political relationship between the state and the established Church in England—a relationship then so close that we can scarcely imagine it today.

However, I am going to pick out for illustration only four interconnected manifestations of the relationship of science and politics. I shall have to hop back and forth in chronology and geography, although Britain will predominate. First, there is an intellectual relationship in terms of theories and ideas. The flow of influence was once mainly from science to politics but evolutionary biology and genetics in the last 100 years have shown an often devastating transfer of ideas back and forth between biology and politics. Second, governments and politicians, as well as scientists themselves, have appreciated that science can be harnessed for national power and pride although there have been occasional and dramatic denials of this. Third, science has been mobilized for national welfare. Fourth—and moving from the abstract to flesh and blood—many scientists, whether as individuals or groups, have been politically very active. With these four aspects of the relationship in mind, and beginning with the seventeenth century and its scientific revolution, I can think of almost no major political movement or upheaval in which science or scientists as people have not been involved.

First the flow of theories and ideas from science to politics. This was most obvious in the three political revolutions—three of the climacterics—of modern history: the American and French revolutions of the 1770s and 1780s and the Russian Revolution of 1917. There were of course many causes and influences behind each of them. Both the American and French revolutions were, however, powerfully influenced by the ideas of the eighteenth-century Enlightenment and, as Isaiah Berlin has said, the entire programme of the Enlightenment, especially in France, was 'consciously founded on Newton's principles and methods and derived its confidence and its vast influence from his spectacular achievements'.[3] Two great men of science—Benjamin Franklin and Thomas Jefferson—were among the Founding Fathers who framed the constitution of the United States, and they were very conscious of the correspondence between their proposed system of constitutional checks and balances and Newton's system of mechanics. From the birth of the Republic Americans put their faith in a combination of democracy and science as a sure foundation for human progress.[4]

Leaping forward to the Russian Revolution of 1917, the Bolsheviks were influenced not so much by a specific set of scientific theories but by the more general belief that the theory which inspired them—Marxism—was scientific, and that Marxism and the natural sciences were linked, because both were materialist and both employed the dialectical method. Later these close theoretical ties between Marxist theory and natural science rebounded on science when they were used as a justification for policing science and declaring some theories to be un-Marxist and hence unscientific.

England's seventeenth-century revolution, a century before those in America and France, was not an international climacteric although England beheaded one king, restored another, and drove out yet another. This succession of revolutionary events was not stimulated by the principles of physics and cosmology, but Charles Webster of Oxford has shown[5] that the social and economic policy of Cromwell's Puritan government was strongly influenced by ideas derived from, and about the role of, science. Religion was to them a cement between science and politics: 'the Puritan Revolution was seen as a period of promise when God would allow science to become the means to bring about a new paradise on earth'. This millenarianism was reinforced by Francis Bacon's belief in the possibility of restoring man's dominion over nature.

And so, apart from—but sometimes including—the giants of the times in abstract science, there was a wide circle of scientists who saw in science not only the power of providence but also the means of social amelioration. Politicians and intellectuals combined in utopian planning and the advancement of learning. One example of the application of science to political purpose was political arithmetic—the first attempt at national accounting, at systematically applying the techniques of mathematics and natural science to government. As Professor Price has emphasized, this Puritan philosophy was carried by the colonists to the New World and was another reason for the strong bond between American science and government.

With the Restoration in England of Charles II in 1660, such ideas about the role of science persisted, albeit under different labels, and they provided the impetus not only for the foundation of the Royal Society but for the use of science in the political purposes of the new regime. This brings me to my second manifestation of the intimacy of science and politics—that is the use of science for enhancing national power. The supreme political objective in Europe was the power of each nation state relative to the power of its neighbours. In the late seventeenth century power became an end in itself, reckoned both in military and economic terms, and trading war was a necessary ally of military war. Intellectual achievement moreover added to the pride and glory of the state. Science was encouraged and conscripted to serve these ends, these wars and to embody this pride. It was natural that the Frenchman Colbert—the supreme advocate and architect of this system—should be chief creator in 1666 of the Paris Academy of Sciences. Since navies and trading fleets were of overriding importance, scientific and technological research in aid of navigation was especially liberally endowed by governments, in particular through observatories.

As the political philosophies of different countries diverged in the eighteenth century, so did the relations between science and government. Thus the traditions of centralized and publicly endowed science and technology survived in France throughout the eighteenth and nineteenth centuries in spite of political upheavals, and until 1830 or so French science flowered more profusely than science in any other nation. Thereafter it suffered from over-centralized control. In England by contrast, as *laissez-faire* doctrines developed and flourished in the eighteenth and early nineteenth centuries, central government was almost totally uninterested and uninvolved in science. Neither the Royal Society nor the universities had public endowments and no major state scientific institution was established in Britain between the Royal Observatory in 1675 and the takeover of the Royal Botanic Gardens from the monarch in 1840.

The doctrine of *laissez-faire* reached its zenith in Britain just as industrialization accelerated into a so-called revolution. Industrial strength became more than ever the foundation of national power, reckoned in dominance over markets and command over resources, which in turn stimulated imperial expansion. Science was involved more than is commonly supposed even with the early industrialization and subsequently, in the late nineteenth century, scientific inventiveness and the diffusion of scientific competence became fundamental—most of all in two key industries on which many others depended: the chemical and electrical industries.

State endowment of scientific research and promotion of scientific education became highly political questions in Britain from 1830 when Charles Babbage, the premature inventor of the computer, published a fierce polemic deploring the decline of science in England and urging government financial support.[6] Other scientists disagreed with him, wanting science to remain, above all, independent of the state. But by the 1870s scientists and engineers were largely united in their pressure for governmental action, as a remarkable series of parliamentary and Royal Commission inquiries seemed to confirm Babbage's gloomy cry 'we are fast dropping behind'.

Britain, hitherto industrially supreme, had been very obviously outclassed in the Paris international exhibition of 1871, most notably by Germany. The princes and dukes of the fragmentary German states had endowed science not directly but by financing a system of universities—a vast organization for intellectual work—where science, alongside the humanities, flourished exceedingly. A parallel system of equally excellent technological universities and mining academies was created and the research spirit spread to industry. So Germany not only began to lead the world in scientific research but also produced for its labour force large numbers of able rank-and-file scientists. Yet despite increasingly urgent warnings, the British government, averse from increasing public expenditure, did almost nothing. In 1885 when the four Scottish universities received a total grant of £30000 a year and the English universities nothing, Prussia was spending nearly half a million pounds a year out of taxation on her universities. Japan, just emerging from feudalism, was likewise pouring money into scientific and technical education. Technical education in Britain received its first substantial cash in 1890 only because a resourceful MP used parlimentary techniques to divert into it the whisky money —that is government money which was intended to eliminate superfluous public houses but which teetotallers and the liquor trade alike rejected.

The clearest sign of relative British decline in science-based industry came in that part of the chemical industry[7] which sounds delightfully peaceful and not all that essential: synthetic dyestuffs. Inspired by the great German chemist Liebig, an English chemist, Perkin, working under a German professor, Hofman, at the Royal College of Chemistry in London made the first brilliant discoveries. But within fifteen years the lead had passed to Germany, to which Hofman and his colleagues had returned, depressed by the unprofessional nature of the British dye industry, the backward state of British organic chemistry and his cramped London quarters. Prussia wooed him back with fabulous laboratories. And so by 1913 on the eve of the First World War Germany produced 135000 tonnes of synthetic dyestuffs out of a total world production of 162 000 tonnes. Britain—the pioneer—produced 5000 tonnes.

I shall return to dyestuffs later, but meanwhile let us note other branches of the chemical industry which were, through the logic of their new science-based processes, becoming involved in commercial arrangements which overrode national power. In the heavy chemical industry the new Belgian Solvay method for producing the key product—soda ash—was the first complex cyclical chemical process. The vast scale of operations led to arrangements for price fixing, market sharing and regulation of production which comprised an elaborate system of industrial diplomacy conducted without reference to governments and stretching into every continent: a new force in international politics.

The third division of the new chemical industry was explosives, which cut across national frontiers and in which peaceful and warlike purposes were entangled together. Here the story centred round that strange Swede Nobel—a very ingenious chemist and a most ruthless businessman, a pioneer of a specially tight form of multinational corporation which was in league with all the other explosive companies in the world. Nobel's first three major inventions—the detonator, dynamite and cordite—were essentially for civilian purposes but his fourth in 1887, was a propellant, ballistite, specifically for gunnery, which it was to revolutionize. The civilian inventions were as important as the military. For they opened up the world—tunnels through mountains; canals, mines and goldfields. Nobel was especially anxious to set up in Britain because of the colonial opportunities—his mouth watered at the thought of Indian railways, Australian and South African mines. Science and politics, that is, were intermingled in imperial expansion.

Naval voyages of exploration had long been linked with scientific investigation, such as the partnership of Captain Cook and Joseph Banks. But scholars are only now beginning to study the full relationship between science and the high noon of the Empire. The popular image of a British Empire created and governed by Oxford Greats men has obscured the pervasive role of those scientists such as the botanists and geologists who, with their professional institutions, were deeply imbued with imperial and economic purpose. It was in India, the brightest jewel of the imperial crown, that the association between science, technology, and imperialism was especially strong.[8] Science was used to exploit her geography and natural resources and then, as the nineteenth century ended, it was also seen as a key ingredient in a new imperial doctrine of 'optimistic liberalism' in which scientific research and method would solve acute practical problems, most notably chronic famine. Attempts to systematize scientific advice to the government of India went far beyond those in the mother country and when members of India's cadre of scientific and engineering civil servants and army officers returned home they were prominent in the agitation for state support for science in Britain. The titles *Imperial* College of Science and Technology of 1908 and, later in 1926, *Imperial* Chemical Industries speak volumes.

I now come to the third manifestation of the intimacy of science and politics—that is, in welfare. In nineteenth century Britain, industrialization and urbanization had created problems which demanded state intervention and expert advice from scientists: safety in mines and in the gas and chemicals (especially the explosives) industries; the problems of the new electricity industry; employment in factories; food adulteration; pollution; sanitation; water supplies and so forth. Scientists, including the most eminent, served on innumerable inquiries and commissions and scientific experts entered public service; they were deeply involved, that is, in politics. Early in the twentieth century science-based state welfare moved ahead, partly favoured by the liberal climate of the day and partly because the Boer War had demonstrated technical inefficiency and such poor health of Army recruits that the phrase 'a C3 nation' was coined. The foundations of the Medical Research Council were laid in the first National Insurance Act of 1911.

What of the United States, which was to be the dominant nation of twentieth-century science? There, science and government had mingled inextricably as the nation expanded westwards into new and strange environments.[9] The federal government supported a great research effort in which military organization and civilian scientists combined and which covered a host of sciences: astronomy, hydrography, oceanography, terrestrial magnetism, meteorology, topography, geology, botany, zoology and anthropology. The parallels with the science of British imperial expansion were the more apparent as the United States Navy, scientists and commerce combined in overseas exploration in Central America, the Middle and Far East. Because the federal government was the major patron of science, science became at times the centre of political storms in the struggles both between federal and state power and between government power and private initiative.

The American Civil War—'the last of the old wars and the first of the new'—pushed the federal government into science-based engineering. During this war the National Academy of Sciences was set up by legislation and an Act provided for land grant colleges from the sale of federal lands. The colleges were a framework for the later wider diffusion of science in the American population and they, with the growth of the great university research institutes, private foundations and industrial research laboratories, were to make the United States the leader of science-based industry and scientific research. As the twentieth century opened, science and politics were joined, as in Britain, in the new era of liberal progressivism. Now that even America's resources seemed finite, federal government science was mobilized to a new fundamental natural problem: conservation.

By the twentieth century science-based industrial strength was necessary not only for market power but for national survival. When the First World War broke out it represented in scale and intensity the totality of scientific progress devoted to destruction. In Britain, the Cassandras who bewailed science policy over nearly 100 years proved right as she found herself ignominiously dependent for crucial science-based products on imports from Germany. The First World War has often been called the chemists' war; so I return to the dyestuffs industry. The near-disappearance of the British industry meant not only a great shortage of dyes for uniforms but the absence of the organic chemical industry to produce other crucial military items from drugs to photographic materials. In 1914 the British government hastily began to rebuild the industry by creating a British Dyestuffs Corporation and endowing its research. Then, realizing the need for much broader action, it set up the Department of Scientific and Industrial Research.

To replace another crucial German imported chemical—acetone—a brilliant immigrant chemist, Chaim Weizmann, was brought from Manchester to London where the government gave him a special laboratory. Lloyd George implied in his War Memoirs that the Balfour Declaration of 1917 which promised the Jews a national home in Palestine was a reward for Weizmann's scientific work. Weizmann denied this but nevertheless believed that his contacts, through this work, with ministers and officials, including Churchill and Balfour as well as Lloyd George, had enabled him to promote the Zionist cause.[10]

In the brief 20 years between the two World Wars science advanced dramatically. But in Fascist countries and Russia the flow of political into scientific theory severely damaged biology. It also stimulated the evil pursuit of racial purity. Fascist countries sacrificed national power to antisemitism and Hitler himself said 'if the dismissal of Jewish scientists means the annihilation of contemporary German science then we shall do without science for a few years'. And so scientists of every discipline left these countries. However, they left in their hundreds rather than their thousands and very many scientists and engineers of the highest quality remained in Germany, often working loyally for the German war machine. The refugee scientists had immense influence in their host countries as Oxford science alone bears witness and, in this confluence of science and politics, the refugee scientists in Britain and the United States changed history. For without them an atomic bomb would almost certainly not have been made before the end of the Second World War and if these physicists had not left Germany, the latter might have made the bomb first.

Before I speak about the atomic bomb, I return briefly to the last of my manifestations of the relationship between science and politics: the active involvement of scientists as individuals in politics. Since a roll-call is impossible, I merely mention a few men popularly associated with science in its purest form. Thus an earlier Herbert Spencer lecturer hailed Francis Bacon as the epitome of, I quote, 'discipline, charity and innocence'[11] overlooking completely his career as a corrupt place-seeking politician, the only Lord Chancellor of England to be impeached and imprisoned. Isaac Newton was, for 30 years, in charge of the Royal Mint, pursuing counterfeiters and clippers with relentless ferocity at a time when the soundness of the coinage was essential to political stability. In the French Revolution many famous scientists were deeply and personally involved, albeit on opposing sides. Lavoisier, the father of modern chemistry, went to the guillotine as a tax collector. Cuvier, the great anatomist and zoologist presided over the Interior Department of the Council of State in the post-Napoleonic monarchy. Pasteur was briefly a French senator and, deeply patriotic, said—how wisely—'*la science n'a pas de patrie mais le savant doit en avoir une*'. In the twentieth century, Einstein, another epitome of pure science, was committed to internationalism, pacificism, socialism and Zionism, and was the centre of political controversy even in the 1920s. Long before Hitler, a German Nobel prizewinner, Lenard, led a group which opposed relativity as part of a vast semitic plot to corrupt the world. And in August 1939 it was Einstein who wrote to President Roosevelt urging him to take action on the new dramatic, scientific possibility—an atomic bomb.

In the last part of the chapter I shall show how the Second World War brought a new relationship between science and politics, in particular through the development of the atomic bomb. For British and American scientists became directly involved in enormous political-cum-moral questions. First, should an atomic bomb be made at all? Second, should the bomb when made be used against Japan? Third, should some attempt be made, while the war was still on, to forestall a postwar nuclear arms race?

First, the decision to make a bomb. Despite Einstein's letter it seemed by early 1940 that for scientific and technological reasons an atomic bomb was impossible. But then in the spring of 1940 two refugee scientists in Birmingham, England—Rudolf Peierls (so well known in Oxford) and Otto Frisch—wrote an extraordinary three-page memorandum showing how and why a bomb was possible.[12] They saw clearly the strategic and moral implications of a weapon so devasting in its explosive and radiation effects. The Peierls-Frisch memorandum drew startlingly simple conclusions from theoretical data that had been published three days before the war. If they were right, it seemed very probable that the scientists left in Nazi Germany would draw the same conclusions. It seemed essential therefore to start production as soon and as rapidly as possible—the more so as Hitler's armies were about to sweep to the Channel ports, leaving Britain virtually alone until Germany attacked Russia in mid-1941 and Japan attacked the United States on December 7 1941. So British and refugee scientists in England worked frantically to see if an atomic bomb was possible and it was their report that pushed the small and floundering American project off the ground. The German scientists' project, after a good start, most mercifully floundered too. The suggestion by some German physicists after the war that they did not make an atomic bomb because of their moral scruples is quite untrue. Russian physicists also made a good start but the German invasion stopped their work.[13]

The British had begun their bomb work on the principle of deterrence, and until the last months of the war they and the Americans believed the Germans were close behind them. But as the huge

American project took shape, the bomb was seen not simply as a deterrent and insurance but as a possible powerful addition to the allied armaments for actually winning the war.

Should the bomb, that is, be used? By the time the first bomb to be made was tested in the summer of 1945 the German war was over. Was it really necessary to use it against the Japanese? This is not the occasion to discuss this decision in detail but it is not true that, as is often alleged, it was taken with little thought. There was much discussion and heartsearching and I believe that while some of those on the atomic project certainly did want their bomb to end the war and while neither America nor Britain wanted Russia's entry into the Japanese war, the prime reason for the release of the bombs was to end the Japanese war with its immense daily slaughter as soon as possible.

For this chapter, the relevant point is the role of the atomic scientists in the decision. It is often believed that the bombs were dropped in the face of their opposition and that their tender consciences had little influence. This also is quite untrue.[14] The scientists were deeply involved in the operational planning. Moreover, the American committee under the Secretary of State for War which was set up in the summer of 1945 to consider the use of the bomb had three famous scientists among its six members—Vannevar Bush, James Conant, Karl Compton—and had besides a scientific advisory panel with four eminent members—Arthur Compton, Enrico Fermi, Ernest Lawrence, and Robert Oppenheimer. Twice during the summer of 1945 the committee and its panel discussed with great earnestness whether the Japanese should be given some striking but harmless demonstration of the bomb before it was used militarily. They believed, however, that no demonstration in a desert area could be made so convincing that it would be likely to stop the war: or the bomb might be a dud; or the Japanese, forewarned, might attack a demonstration plane; or bring prisoners-of-war to the test area. If the demonstration failed to bring surrender the chance of administering the maximum surprise shock would be lost. There were no bombs to waste and it was vital that a sufficient effect be quickly obtained with the few there were. Viewing war, not the bomb, as the fundamental problem these scientists thought a military demonstration might be the best way of furthering the cause of peace.

The committee and its scientific panel held to these opinions in spite of the well-known memorandum by six Chicago scientists headed by Nobel prizewinner James Franck. This was a powerful plea that hopes of international control of atomic energy would be fatally prejudiced if the United States released this new means of indiscriminate destruction upon mankind. Nevertheless even the Chicago scientists added the rider that, if the chances of international control seemed slight, not only the use of the bomb against Japan but even its early demonstration might, on secrecy grounds be contrary to the interests of the United States. When a poll was held among the scientists in the Chicago laboratory, over 60 per cent favoured a military demonstration and for various reasons the percentage at other laboratories in the project, if they had been polled, would almost certainly have been higher. As the scientific panel tendered the advice that they saw no acceptable alternative to direct military use, they added an important disclaimer.

It is clear [I quote] that we, as scientific men, have no proprietary rights. It is true that we are among the few citizens who have had occasion to give thoughtful consideration to these problems during the past few years. We have, however, no claim to special competence in solving the political, social and military problems which are presented by the advent of atomic power.

The top political leaders of America and Britain had difficulty in understanding that atomic bombs were different in kind, as well as in size, from their predecessors, and that the bomb could not be kept secret after the war. They were also reluctant to face the third big wartime political question about the atomic bomb: what could be done to forestall a postwar nuclear arms race? Attempts to make the politicians understand the problem were made most of all by Niels Bohr, the atomic physicist, one of the scientific giants of the century.

In 1943 the British brought Bohr out of occupied Denmark in the best cloak-and-dagger style and informed him about the atomic bomb.[15] He realized at once that this would lead to still more terrible weapons such as the hydrogen bomb. Above all he was overwhelmed by the realization that this weapon of unparalleled power must bring fundamental change to the world. Bohr was thought to be the most unworldly of scientists, but at a time—1943—when most people in the West were talking about their heroic Russian allies, he saw that postwar life would be dominated by tension between Russia and the West, and that the only chance of forestalling a nuclear arms race between them was to tell the Russians about the bomb before it had been used and to attempt to agree on control before there was any threat of duress. This belief became the main preoccupation of Bohr's life for the rest of the war and he spent his time in political antechambers tirelessly advocating it. Churchill and Roosevelt were desperately worried about Russian behaviour in Europe at this time and, partly as a result of this, Churchill especially had deep suspicions of Bohr. He and Roosevelt signed an agreement, one clause of which implied that Bohr might be a Russian spy.[16] Churchill wrote 'It seems to me Bohr ought to be confined or at any rate made to see he is very near the edge of mortal crimes'.[17]

Bohr's efforts have been described by a political scientist as a 'flight into the higher mysticism away from the unpleasant and unacceptable world of politics'.[18] But Bohr's proposal was essentially practical. He knew that the Russian physicists were extremely good and that once a bomb was dropped there could be no secret. To inform Russia officially would therefore carry very little risk and might conceivably bring benefits. Not to inform Russia would bring no benefit and would carry the risk of intensifying suspicions. Bohr's idealism, that is, was set in a very practical framework of limited objectives as he looked to a future when civilized life might be destroyed in a flash. If Russia had been formally consulted about the bomb during the war—she knew of it from spies and had already begun her own project—it might have made no difference. The fact that she was not, guaranteed that the attempts made just after the war to establish international control, which might have failed anyway, were doomed.

After the war, American atomic scientists were up to their necks in politics and many of them were powerful within government. One eminent physicist wrote about 'the Perils of being Important.'[19]

> suddenly physicists were exhibited as lions at Washington tea parties, were invited to conventions of Social Scientists, where their opinions on society were respectfully listened to by lifelong experts in the field, attended conventions of religious orders and discoursed on theology, were asked to endorse plans for world government and to give simplified lectures on the nucleus to Congressional committees.

For a decade or so there was scientific solidarity and symbiosis between scientists and government. The end of this era was signified by the bitter government hearings of 1954 about Oppenheimer's security status.[20] In these hearings Von Neumann the great mathematician said:

> All of us in the war years . . . in scientific and technical occupations got suddenly in contact with a universe we had not known before . . . we were all little children with respect to the situation which had developed, namely that we suddenly were dealing with something with which one could blow up the world . . . we had to make our rationalization and our code of conduct as we went along.

The scientists' rationalization and advice which looks most wise in retrospect is that which combined a hope for mankind, an anxiety to restrain the technological exuberance of the arms race with a severely practical appraisal of the risks and benefits of a particular course of action. This was certainly true of Bohr's proposals.

The postwar story of the scientists and nuclear weapons was to emphasize the indivisibility of scientific skills and political values and to query the innate superiority of the innately rational scientist over the politician. The scientists of the atomic era indeed became acutely conscious of phenomena which rule political life: the conflict of desires and aims, the conflict between the interests of different generations, the difficulty of calculating consequences. In the years of their ascendancy they proved that they were not all-wise nor indeed all-wicked but infinitely human. They could change their minds with devastating speed. They could be both wise and foolish, both myopic and far-sighted, both judicious and ridiculous, both clear-headed and muddled. They turned out to be, indeed remarkably like the politicians.

References and Notes

1. Bernal, J. D. *The social function of science.* Routledge, London (1939).
2. Ravetz, J. R. *Scientific knowledge and its social problems.* Clarendon Press, Oxford (1971).
3. Berlin, Isaiah, 'Einstein and Israel' in *Albert Einstein: historical and cultural perspectives* (eds. Gerald Holton and Yehuda Elkana). Princeton University Press, Princeton, NJ (1982).
4. Price, Don K. *The scientific estate.* Harvard University Press, Cambridge, Mass. (1965).
5. Webster, Charles *The great instauration.* Duckworth, London (1975).
6. Babbage, Charles *Reflections on the decline of science in England.* B. Fellowes, London (1830).
7. For all the paragraphs about the chemical industry see especially Reader, W. J. *ICI A history Vol. 1* Oxford University Press, Oxford (1970).

8. See MacLeod, Roy, 'Scientific advice for British India, 1898-1923'. *Modern Asian Studies*, **9**, 3 (1975); and Dionne, Russell and MacLeod, Roy, 'Science and policy in British India, 1858-1914'. *Colloques Internationaux du CNRS*, no. 582.

9. Dupree, A. Hunter *Science in the federal government.* Harvard University Press, Cambridge, Mass. (1957).

10. Weizmann, Chaim *Trial and Error.* Hamish Hamilton, London (1949).

11. Ravetz, J. R., '. . . et augebitur scientia', in *Problems of Scientific Revolution* (ed. Rom Harré) Clarendon Press, Oxford (1975).

12. Gowing, Margaret *Britain and atomic energy 1939-1945.* Macmillan, London (1964).

13. Holloway, David, 'The Soviet decision to build the atomic bomb 1939-1945', *Social Studies of Science*, **2**, 159-90 (1981).

14. Hewlett, Richard G. and Anderson, Oscar *The new world.* Pennsylvania State University Press (1962).

15. Gowing, *Britain and atomic energy 1939-1945.*

16. Ibid. Appendix 8.

17. Public Record Office PREM 3/139-8A, 298-9.

18. Watt, Donald C. 'The historiography of nuclear diplomacy', *Science*, **194**, 174-5 (1976).

19. Samuel K. Allison, 'The state of physics or the perils of being important', *Bulletin of the Atomic Scientists*, January (1950).

20. United States Atomic Energy Commission. In *The matter of J. Robert Oppenheimer.* United States Government Printing Officer (1954).

5 *The responsibilities of scientists*

H. MAIER—LEIBNITZ

Professor Emeritus of Technical Physics,
The Technical University, Munich, Germany

Like many of those who have been invited to deliver a Herbert Spencer Lecture, I feel worried. Being a physicist who has spent most of his time trying out new experimental methods for investigating the properties of matter, not only can I contribute nothing to philosophy but I cannot even express myself in terms which a philosopher might find appropriate. Therefore, I was very glad to read that Einstein, in his Herbert Spencer Lecture in 1933, said: 'If you want to find out anything from the theoretical physicists about the methods they use, I advise you to stick closely to one principle: Don't listen to their words, fix your attention on their deeds'. The fact that he gave so many speeches in spite of this probably has to do with a feeling of responsibility that we shall have to discuss later. As for myself, I feel I must restrict my report to what I have felt myself, what I have seen with my colleagues, what I have tried to achieve, and what I try to fight or to spread.

We all have duties which we have accepted, which have been assigned to us, or which we have chosen. Mine are '*Die technische Physik in Forschung und Lehre zu vertreten*', to promote technical physics in the domains of research and by teaching. Such a general definition includes nearly all the activities we might choose. Our responsibilities will thus arise from our activities and our duties, from the choices we make, the decisions we take, from what we do not do, and from consequences of actions which we have not foreseen. As long as we only do what everybody expects us to do, the problem of responsibility will rarely come up. However, in contrast to many other professions, the normal duties of a scientist can never be completely defined from above or from the outside. Much must be left to the scientists themselves, and they even have difficulties in explaining what they do. This is one reason why the discussions on the responsibilities of a scientist are often confused. Another difficulty lies in an uncertainty about the meaning of the word responsibility which is often distorted for reasons of rhetoric. We shall define it as we proceed in our discussion.

The responsibility just mentioned, which is connected with our normal duties, is necessarily rather vague in nature. For a scientist —and not only for him but for an important minority of the population—his profession is a vocation, is what he likes most to do, what he thinks he must do, what only he can do. A true scientist always does more than others may demand of him, more in kind, not necessarily in quantity. A physicist, so I have learnt from my peers, is a person who, after working for two years in a given field, knows more about something, can do something better than anybody in the world. (This, by the way, might be a good criterion for a young scientist who has to decide whether or not to continue a career in research.)

As a consequence, nobody can give orders to a scientist on many important points. There, he must have freedom to decide himself. Freedom, however, always means responsibility, too, a responsibility which we have to choose ourselves, alone or together with our colleagues.

Let me give a few examples from the field of research, to begin with. When starting out, a scientist will have certain ideas and aims which may be his own or may have been suggested to him by some programme or person. If a physicist aims at progress in the knowledge of chemical binding, he may, after a study of the literature, decide to investigate the distribution of electrons in crystals. Here, previous attempts by others have been rather unsuccessful, so he tries something new: using gamma-rays instead of the usual X-rays. He thinks he can prove that this will help, but for a long time nobody will believe him, or even seriously listen to him. Only by his evident success can he convince anybody. This is an example of a general aversion of nearly everybody towards things that have never been done, and certainly of the impossibility of having them planned by outsiders.

Another example is cancer research on which billions have been spent world-wide. However, the causes of cancer are largely unknown, so progress must come much more from unexpected discoveries than from systematic search as in programmes. The responsibility of the researcher consists in keeping in mind his general aim, not getting lost in tracks which lead nowhere, but doing research as best he can and keeping his eyes open for unexpected results.

Another example is teaching. In my country, we now have a wave of regulation at our universities, prescribing hours, types, and content of teaching, regardless of the persons who teach and of the contributions which only they can make. This, if it must be taken seriously, leads to a waste of the time of the best scientists, and to a standstill in the kind of teaching which lives by constant renewal. One great task has arisen through the ever-increasing speed with which new knowledge arrives nowadays. We need critcal analyses of the new data, and then their presentation in the simplest form possible without losing their content. We need a constant revision of the scientific language. Otherwise it will become increasingly impossible for coming generations to start their work on the basis of what is known. And we need the best possible textbooks which enable the specialist and the student to think in terms of more than a very narrow field. These are examples of what the best scientists might consider their own, self-chosen responsibility.

We shall pass relatively quickly over a responsibility which is familiar from other fields than science. Nobody must do damage to persons or property in the course of experiments or produce a situation which may become dangerous later, or he will be threatened by sanctions. A researcher may not cause explosions, or poisonous smoke, or the pollution of water. In clinical medicine, he has to follow the ethical rules of the medical profession.

A scientist will feel the responsibility within himself in such cases. In most cases it will coincide with the opinion of the majority, or with standards like the Hippocratic Oath which have been valid for a long time. In many other cases, rules of safety and of correct conduct have been set up by law or by government agencies which have to permit or forbid certain activities. However, difficulties arise in borderline cases, or when new situations come up which were not foreseen in earlier agreements or regulations. The most striking example was what was at first called 'gene manipulation', by which organisms may be created which had not existed in the world before. The scientists themselves were frightened by the idea that such organisms might enter a world in which no enemies existed which would keep them at bay. So they suggested extensive measures against such dangers, and legislation in some countries followed suit. But then, with increasing use of the method which promised very important and useful applications, they found that the new organisms proved unable to survive outside the laboratory because evidently they had not been adapted by evolution like the naturally occurring organisms. The conclusion that therefore there is no danger is of course a conclusion from n to $n + 1$, but indeed most restrictions have since been lifted, and a great development is underway world-wide. This is a striking example of how, in a new situation, the scientists have a responsibility that nobody else can take away from them.

It should be noted that this type of responsibility includes future damage beyond the lifetime of those who act today. A dam must be safe for a long time, a dump for chemical waste must not liberate poisons later. Our responsibility for future generations which is so eloquently advocated in the book of the philosopher Hans Jonas[1] is familiar to the legal system of today. It must be admitted, however, that many cases of possible future damage are still far from being considered susceptible to sanctions. Examples are the changes of climate due to the destruction of the woods, or the exhaustion of the world's reserves in fossil energy or raw materials, let alone the more imaginative hypotheses like the degeneration of mankind through genetic manipulation.

We now come to the responsibility of the scientists (and to a degree we must include here the engineers as 'applied scientists'), for what use is made of their discoveries and their work in general. The fundamental point is that the scientists do not, or do not alone decide on application. They share, though not always, their responsibility with industrialists or with politicians. This is evident, and our leading politicians have stated it very clearly,[2] but only too often it is forgotten in public discussions. Taking this point into consideration, we shall now try to analyse more closely the responsibilities of the scientists for the consquences of their discoveries.

A first point—and it is often raised—is that many evils in the world would not exist if there had not been any scientists to create a basis for them by their discoveries. The extreme wish would be that science would not exist which, of course, is quite unrealistic in the present world. However, it is interesting to consider this hypothetical case for a moment in order to see whether a world without science could be desirable. In the world as it is, it would not be desirable but catastrophic to try to live without science. Due to the industrial revolution, and to an increasing extent due to the progress made possible by science, the world now has such a large and growing population that every effort including more, not less science is needed to nourish them, to fight unwanted consequences of large growth, and to let those parts of the world where life conditions are still much harder than elsewhere, get nearer to a life without hunger and undue hardships. We said above that a scientist has within him a feeling of responsibility, for instance for having chosen research as his profession. This feeling cannot tell him that science is universally bad and that he and everybody else should abstain from it. This may sound trivial but we must be aware that a fraction of the population, even if it is a small fraction, thinks differently.

Can a scientist as a person or as a member of a group feel responsible for the application of his or their discoveries, even if these discoveries were quite unexpected and nobody would at the time have thought of the kind of application that afterwards became possible? This of course is the case of the discovery of fission by Hahn and Strassmann. Everybody will understand that Hahn was very unhappy about the later development; and another nuclear physicist has said that he feels responsible for his contribution to it just as he feels responsible for his children long after he has lost any influence on their development. However, I feel that the word responsibility does not fit here. Work in science, the humanities and the fine arts belongs to the noblest activities of mankind. We scientists must live with the uncertainty about what will become of our discoveries. You do not ask an artist—at least not in our part of the world—not to paint or write what he feels is an essential part of him, of him as part of humanity. And the same is true for a historian, and must be true of a scientist. Application is a much later stage, and it cannot motivate the scientist on his path to new knowledge except in the vaguest way, making him aware of the fact that scientific discoveries may have consequences. There is overwhelming evidence of useful consequences, so we cannot blame ourselves for what we have done even if we have every reason to regret some of the consequences and have a feeling of guilt and remorse ('*ihr lasst den Armen schuldig werden, dann überlasst ihr ihn der Pein*').

These are things which concern the scientist alone, his conscience, his feeling of responsibility. No reproach can be raised against him by outsiders for the consequences of his unexpected discoveries. And he cannot be made responsible for the existence of the whole class of scientists. Here again is an error which is often committed in public discussions: a single scientist, or a group of scientists in one country or even in several countries, cannot prevent a development which they do not want to happen, by abstaining from research in a certain field, as long as there are strong tendencies in another part of the world to promote this development, maybe for national or selfish purposes. A single scientist, even of the highest class, cannot even cause an important delay in the course of science. We have looked at a number of important discoveries, and must believe that without the original discoverer they would have been made some three or five or, at the most, ten years later. This should make both us and our critics a little more modest in the assessment of the importance of our individual contributions. It is true that sometimes we find or invent things that could have been found 20 or 30 years earlier, like the laser or maybe the Mössbauer effect, but at such an early time only a great genius could have acted alone, and in a climate of opinion that would have isolated him before he started.

In this connection, I should quote a sentence which was pronounced by a German member of parliament during a televised broadcast in Austria with Edward Teller as a participant: 'We shall not tolerate that you, with your ideas and inventions, dominate our future . . .'. This has to do with what we have just said but there is something else: the member of parliament, and there are many like him, seems to think that we try to dominate the world, or at least that we are an important obstacle against a better form of society. We tend to regard this as ridiculous, but nothing is so ridiculous if many people believe it. Is there something like a conspiracy of the scientists, and are there examples in the past?

If we think of the movement that started with Copernicus and Galileo and went on with Kepler and Newton to Voltaire and Diderot, we see indeed a movement that went against some of the ruling ideas of their time, and had great consequences. This is one side of the picture. The other side is that even if or because the world is full of conspiracies, calling an idea, an attitude of a group or simply the group a conspiracy is a well-known weapon in political disputes. Hitler did it with the Jews, and there are many examples today wherever there are conflicts; Iran is just one such example. Whenever a group has international coherence (which is very much the case with the scientists), the conspiracy is considered even more dangerous. If we take the nuclear scientists as an example, there are many people who wish and believe that the scientists should form an international conspiracy against the production of nuclear weapons or nuclear reactors; and there are those like our member of parliament who think they should be prevented from forming a conspiracy for technological progress.

The real point is that the entire discussion is concerned with the application of scientific results, not with philosophy of science as it was in the period from Copernicus to Diderot. And here we must be aware that there are two kinds of research: that with the aim of improving our knowledge of the world, and that aimed at finding answers to questions which come up in connection with definite applications. Such research evidently is already part of the process of application and we must treat it as such even if we know that the distinction between pure and applied research is often quite ambiguous. A scientist may, consciously or not, do research which serves or prepares definite applications.

Here again, we must come back to what we have said earlier: decisions on the application of results are not made by the scientists, and their activity must be seen together with those of governments and industry, and with the influence of public opinion. The responsibility is a shared one, and the scientist is included in it not only as a contributor of scientific knowledge but as a citizen like other citizens. This has important consequences which are too rarely made clear in public discussion; a scientist must have the right to join a political group or party, or to have an opinion about the government or the constitution and to act accordingly, thus sharing his responsibility with many others. If he likes his government and the government says that defence is of vital importance for the country and for the world, because without a capacity for defence there will be no peace, then he must feel free, in the name of his government, to join a research effort for defence. Of course, he can still say: 'I don't like defence research and I shall not take part in it', and in a country like ours he will be free to do that. But he must know that this is the attitude, for example, of the steak-lover who does not like the butcher but certainly cannot blame him. We must distinguish between the personal decision of a scientist to take or not to take part in research for certain applications and the declaration of a general responsibility for these applications. The latter must be shared by all those who take part in the decision to develop them. Hitting the scientist alone is motivated by the fact that without him the work could not be done; the public considers him as a symbol of progress, and therefore is inclined to regard him as a prime suspect.

If a scientist does not believe in the defence policy of his government he may adhere to, or share in this respect the programme of a party or group that opposes the government. Again he will not be alone or bear the responsibility all by himself. He will contribute his special knowledge, but will act as a citizen among other citizens. The contribution of a scientist may be the same as that of any other citizen but it is clear that there are contributions which only he can make because of his special experience and training, and because only he can do the additional research which may be needed for it to be applied. This gives the scientist a responsibility which is typical for him and which therefore he should not avoid; he must contribute everything to the application, or more generally to the common cause, that only he can contribute. When we say 'he must' this evidently cannot mean that every scientist must do that. Science would be lost if the best scientists could no longer spend most of their time doing good research. 'He must' evidently means that a small but competent minority must give enough time for such tasks. And when we say 'everything', we do not mean that a scientist should contribute even if his conviction as a citizen forbids him to do that. He must speak out for an application or work for it. Or he may warn against it. When he speaks as a scientist, however, he must not violate his professional standards of truthfulness and honesty.

The fact that scientists act as citizens is clearly visible by their participation in many movements, including those that are directed against science and against progress. Indeed, as far as I know, all the important warnings of dangers that might arise from applications like nuclear energy, pollution, carcinogenic substances and so on, have first come from the scientists themselves. This includes the reponsibility for future generations but it also means participation in the discussions about the needs of the present generation compared with dangers for the future. This is a good example of something that is generally true; most decisions cannot be made on the basis of technical information only, and most technical information is not purely technical in the strictest sense, comparing only quanitities of equal things. Nearly always it is necessary to evaluate information, and this means that the decisions are political decisions, even if the participants and especially the scientists are not explicitly aware of it.

This is an important point, so let us give an example. Risk assessments for nuclear energy have indicated that if full-scale nuclear energy were introduced world-wide, one might expect that, as a result of general pollution or of reactor accidents the death rate from cancer might increase by one ten-thousandth. In Germany, the death rate from cancer is about 150000 per year, so this would mean 15 additional deaths per year. This figure may be too low or, more probably, too high but that is not the point here. We must compare it with some of the advantages of nuclear energy. If we assume (another uncertain figure) that nuclear energy compared with other energy sources might mean a saving of £200000000 a year in work effort or money, meaning that an effort to this extent could be made for another useful purpose, what does this mean for a comparison? Does it mean that we put a value of £13000000 on each human life, and are those right who say that the loss of a life can never be compensated by any amount of money? Or is it a fact that a higher standard of living in a country, with more research, more help to developing countries, better medical care, will save many more lives each year than the fifteen we have mentioned? We can see that there is always uncertainty, and therefore, and in many cases for other stringent reasons, we need decisions of a political nature.

Before, finally, we deal with the relations between the scientists and other groups of society, we must deal with the ethical standards that govern the work of scientists, standards which once were generally accepted but are now so often neglected that one must speak of a new responsibility to keep them. About 50 years ago, a scientist working in a certain special field was supposed to know all the literature that concerned his work. He was not supposed to make any new statements of his own without quoting all his sources and making sure that nobody had found his results before him.

We are all aware that with the present abundance of scientific information this is no longer possible in every case. This may have somewhat lowered our general standards but at any rate the principle is still universally accepted. Another change has occurred in the same period, and it is more drastic; in the published presentation of results, the methods that were employed and the data that were obtained, were to be described so exhaustively that another competent scientist could completely follow and if necessary repeat the experiment. Nowadays, hardly any journal can afford to publish such complete information. In addition, the slogan 'publish or perish' is no empty phrase. Too many publications are more or less preliminary as regards the results, their interpretation, or their meaning in the context of other work. Thus everybody has learned to take scientific publications less seriously than they had to be taken at earlier times. And all this may have contributed to lower the most important of the ethical standards; a scientist has to make every possible effort to find and test every possible objection that might put his results or theories in doubt. This is still the most important standard for a scientist, quite typical of his work in contrast to most other professions, but we must admit that it is no more quite generally adhered to and that a scientist who violates it might no longer be automatically ostracized. It seems to me that this is a danger signal, and I want to emphasize again the importance of this kind of responsibility.

This lecture is not a scientific publication, and most of what I have said so far has been said by others in some connection. In particular, I think I should mention Michael Polanyi who was among the first scientists to make an effort in this field. But what we do resembles more an amateurish self-portrait than a new sociology of science which would accumulate knowledge so that the next generation can use what the previous one has learnt. What I have to say now is even more scattered and widely uncertain. Both society and the scientists have done relatively little to understand their mutual relations even if we agree that it has become increasingly important to understand these relations.

My first subject is the relation between the two cultures, as C. P. Snow has said. We all know that there is a deep and regrettable separation between those who had a training in science and those, much more numerous, who have a basis in the humanities in the widest sense, and the liberal arts. This separation is of old standing. Leibniz was probably the last who really belonged to both sides, and Voltaire was one of the first who unsuccessfully tried to work actively in physics. In Germany, the separation became very final during the nineteenth century, and it has now pervaded our educational system, and thus our daily lives. Hardly anybody who is not a scientist knows the difference between mechanical force and energy. Decisions of the greatest importance for the economy must be taken by persons who do not know that energy is never lost but may lose its usefulness. This is one side but I think the other side of the picture is more important: We, the scientists, have become one-sided specialists. It is not only that few of us still learn Latin and Greek and thus share a traditional source of communication with others. But we learn next to nothing about philosophy, literature and the arts, about law, economy, psychology, sociology, religion. This has many bad consequences. The worst is that we have lost the basis for understanding values, especially moral values. This would not be so bad if there were a common feeling and understanding between the two groups which enables them to share knowledge and attitudes. But this understanding is badly disturbed, partly because there is a tendency today to belittle moral values in general.

Of course, both sides should do something about this situation. Patient and slow changes in our educational system may help to diminish the knowledge gaps between the two groups. In Germany, we have a movement towards more teaching in mathematics and science. However, I feel this could be a one-sided effort, and besides, what we need is not more but better teaching. We must learn to use basic laws in our thinking and not be filled with more and more facts from science. Better teaching in this sense is a responsibility the scientists must assume. But on the other hand, I feel that it is our responsibility to learn more from the other side. This is useful because it will make us better citizens, and it is much easier than it is for the other side to learn more about science. And there is another responsibility, this time one which both sides can take: to bridge the intellectual and human gaps existing between the two cultures.

We must now consider the relations scientists have with the various groups of society: industry and trade, politics and administration, the public and mass media. I shall not speak of the first point, the interaction of the scientist with those who run our economy. I should think that this interaction is rather satisfactory if we admit, as a scientist will always do, that everything could be better and should be improved. So our first subject must be the world of politics and administration. We shall concentrate on the politician as the prototype in this field. The adminsitration is steeped in the methods and habits of politics and can to a large extent be understood from there, except for the important domain of the bureaucracy with which we need not deal at length here.

The differences between a scientist and a politician seem to be of a fundamental nature, and they can explain most of the difficulties the two groups have with each other. In Germany, a scientist rarely becomes a politician, and those (including some engineers) who have become parliamentarians or cabinet ministers, if they do not belong to the majority who were outright failures, have enjoyed only moderate success. There seem to exist great differences between countries. We once had a discussion among presidents of research organizations from various countries. Some believed that very different qualities are needed to become a good scientist from those needed to become a good politician. Some felt that a scientist cannot but be repelled by the methods that a politician has to use in order to be successful. Only the Frenchman felt that the qualities needed to be a good scientist and a good politician are independent qualities and may only by chance coincide in the same person. He himself appeared to be an excellent example.

What a scientist finds most strange in the behaviour of a politician and what he finds himself incapable of doing, might be summarized as an adaptation to the surroundings in which the politician has to live. In the unfriendly world around him, the politician must never forget to protect himself and his political friends. A responsible member of the government may be convinced that the economy needs nuclear energy but he will not say so in public except in the vaguest terms or when something definite must be decided. He will make large detours and even admit damage to the cause at hand in order to cause damage to his opponents; in the end, however, this may turn out to be useful because it strengthens his position. He will not do even important things unless he can voice arguments for them that may appear misleading to an expert but will impress the public. He will think of subtleties in public issues which have never occurred to the scientists who may consider them unimportant. To the scientist, the whole idea of a political fight with its distortion of arguments is a nightmare, with the suppression of facts which the other side might exploit, with its purposefully false accusations and insinuations. The scientist does not know that without some degree of such behaviour, no politician would ever get near a position where he could promote good decisions.

This is only a small selection of what a scientist might find puzzling or objectionable in the behaviour of a politician. However, a successful relationship between them is utterly impossible if both sides do not have some understanding for each other. We might wish that the politicians showed more understanding for us. But this is not possible on the technical level, and in human matters the politician is, in any case, accustomed to deal with persons of all kinds. So the main task of learning about the partner rests with the scientists. This is a great responsibility, and it is not borne well.

The counterpart to a better understanding of politicians by the scientists is a better understanding of the scientist by himself. The scientist is at a disadvantage because his work keeps him away from the vicissitudes of daily life. He does not learn, as every farmer learns, the unwritten standards of human conduct and human standards. He does not know how to adapt his pure ideas to the necessities without which the best ideas will not only be useless but get him into trouble. The mismatch between his ideas and their acceptance by those who are responsible for making decisions is liable to create within him a feeling of righteousness which makes everything worse. He lacks, as we have said, the subtleness of the politician, and therefore he has a tendency to find, by solitary thinking, 'the' solution of a problem and will never understand why it is not accepted. Part of the difficulty is caused by a fundamental quality of scientific thinking: a scientist is not accustomed to fight for a cause like a politician, who must try to win rather than to be right because otherwise he would never come to useful decisions. For a scientist this is anathema, and so in all political discussions the climate is spoiled for him. Last but not least: due to his preoccupation with rational thinking and his distance from the daily world with the values it places on our actions, he is relatively helpless whenever moral issues beyond those that determine his scientific conduct, become important. This is liable to lose him much of the sympathy that he may have won by his expert knowledge or his undoubtable sincerity. I am speaking here from experience. Our responsibility must be to be more aware of our own limitations, and of course to learn, learn, learn.

The relationship of the scientist with the public follows a similar pattern. Again, there is the problem of mutual understanding, and then of learning from each other. I shall not speak about understanding the public because this really means being a citizen or, if we want to learn more, developing social psychology which might be a fascinating and very useful science. But we must understand what the public expects from us, how it sees us, what it hears when we say something. A clear majority of the population expects more wellbeing of humanity from the work of scientists and is willing to trust their good will and honesty. However, in the fight for power which is part of public opinion, there is now a movement to discredit the work of scientists, to distrust them, to say they are under the influence of industry or of capitalist society. And even those who do not go as far as that are under the impression that it is useles to listen to scientists because whatever problem is publicly discussed, one will find scientists as supporters of both sides. This may have done more to discredit scientists than anything else. There has always been the problem that the public could not understand technical or scientific discussions. One had to trust the scientists for the truth of their technical statements. Now, with the general distrust growing, a small minority can raise objections which seriously affect the credibility of scientists.

The mass media will mirror this development, and unavoidably they must have the greatest influence on it because hardly anybody gets any information about science except through the media. Therefore, if the scientists have a responsibility to inform the public, their first duty is to learn about the mass media, the rules which govern them, and the people who work in the media, above all the journalists and editors with whom they will have to deal. Of course they share this need of learning about the media, about the science of mass communication, with all those who have a public role to play. This science of mass communication is rapidly developing at present, and scientists could learn it as fast as anybody and maybe even contribute to it. So here is another responsibility. I should think that this responsibility is easier to accept than the more general one towards the public, because the number of editors and journalists is limited and we can talk to them, hoping for progress through continuous interaction.

Problems with the media are by no means trivial. I shall give one example. Recently, two American sociologists[3] have asked scientists and journalists about nuclear energy. From the questions, they constructed a 'nuclear support scale' ranging from +9 to −9. From nuclear experts (7.9), energy experts (5.1), to all scientists (3.3), the answers went down: science journalists (1.3), below them prestige press journalists, then science journalists of leading journals and television, and finally, with negative values on the scale, television reporters and producers (−1.9) and public television journalists (−3.2). This is quite puzzling, especially when the authors add that they asked the leading journalists a large number of social and political questions and found that the best predictor of opposition to nuclear energy was the belief that American society is unjust.

The scientist has an important role to play in his interaction with this public and the media. He must—and can, if he tries—understand the wishes of the public and show that he is not a stranger in society. Presenting the result is an important task. There are a few great masters who have found a language which is attractive and convincing even if proof cannot be given without using technical argumentation. There should be more of them because there is a real and general interest to learn how science is expanding our knowledge of the world, our thinking, and our possibilities to act. Journalists can be of enormous help here, and everything should be done to inform and aid them. Journalists and scientists, by their common aim seeking truths, should be natural friends everywhere. It is sad to see that, instead, there is so much muddled misunderstanding in so many places. Even if it is true that scientific news is often distorted in the media, it is no use blaming journalists as a group, especially if scientists do not produce better information.

Another task is to restore some of the confidence which scientists once enjoyed. What are the obstacles? Or to put it another way, what arguments have been successful in the campaign against the scientists and must now be fought? One argument is that scientists are selfish and promote their own interests rather than the common cause. Herbert Spencer would have supported this by saying that everybody is selfish. Indeed, this is hard to deny, and scientists might claim the right to be like everybody else. However, all my experience makes me believe that this is not the whole story. I have seen countless examples of unselfish behaviour of scientists: long, tedious work on ungrateful subjects, untiring efforts in teaching, enormous work in the peer review system, and expecially loving care for the younger scientists who will eventually replace and surpass them. I am inclined to believe that unselfishness is a rightful claim for a scientist though, of course, he can never boast of it because that would make him better than his fellow-citizen. The public, I think, is rather sensitive to this, and ready to honour it.

Another reproach to be dealt with is that scientists, like most people, cannot be expected to stand firm by their established results and consequent convictions but are liable to yield to pressures from their employers or from interest groups, or that they yield, as nearly everybody does, to the very strong pressure of public opinion. This can never be excluded in individual cases, and often it may not even be a conscious action. However, it must be said that it runs strictly counter to the professional ethics of the scientists, and so they will not easily yield to such temptations, less indeed than most other groups of the population.

Something which is not fundamental but is the cause of much damage is a lack of involvement or seriousness of scientists in matters which detract them from their main work in research. This is a great handicap in committees and in public discussions. A scientist will often speak with great conviction about his own field without, however, having analysed all the pertinent data and possible objections.

This is even worse, and it is much worse if the discussion goes beyond the scientist's own field. It has often been observed that in such cases he will transfer his feeling of competence to the other field without any good reason. He will then be a true nuisance, and we must fight him as well as we can.

Still another task for the scientist in public discussions is to fight for the broad view when all the participants are specialists in narrow fields. What is needed is a new type of generalist who, after having proved himself in some speciality, uses his experience to understand something from other fields. Such generalists exist. A physicist, for instance, can understand most problems that arise in engineering, and a psychologist may understand problems in ergonomy.

The last point, and this has already been said, is that the scientist should understand more about life in general and about people. We now have again a list of responsibilities for the scientist, of things he should do beyond his normal duties in order to really contribute what only he can contribute.

I have dealt with such responsibilities at length because I feel that a dramatic situation has developed quite recently, a situation that will require all the goodwill and all the skills of scientists, in addition to efforts by many others. (You will hear very little from me about the possible responsibilities of other groups in society because

I am a scientist myself, and before telling other people what to do I should stick to my colleagues.) Why do I use the word 'dramatic'? I was frightened, when I read recently in a work by Northrop Frye '. . . the "two cultures" situation that Lord Snow misrepresented so grossly . . . was a protest in the name of human concern for survival and freedom against . . . a death impulse in the human mind . . . trying to get control of science and technology'.[4] Are we that far separated in our thinking?

It is true that science and technology is evermore increasing. It is true that the rate of change in our way of life now greatly exceeds the rate of generational change. Unforeseen consequences of technical developments occur more and more often. Periods are too short for the usual and beneficial evolution by trial and error, so its products are weak and fallible. It is only natural that the people who have to live with all that and who are supposed to control it, are frightened, like victims, not like masters of the situation.

The uneasiness of the population focuses on a number of themes which have become symbols even if not all of them are the most important problems: first, the atomic bomb which is the mother of all fundamental fears of our time; and second, atomic energy with fewer good reasons if it is not taken as a symbol for big technology and waste of energy; third, vanishing resources and the warnings of the Club of Rome (strangely enough, the catastrophic growth of the world population appears as a less fundamental worry); and finally technical innovations like microelectronics or gene manipulation. Most of these problems are such that the individual feels he cannot change anything and is helpless.

With all the problems, experts are needed more than ever, and there is a near-consensus that the situation requires more research than we have now rather than less. However, as has been said already, practically every problem has not only a technical side but concerns values of some kind. So it eludes the competence of scientists alone. This situation might be considered as one of the justifications if not also one of the reasons for the recent growth of new movements which encompass much more than the technological problems of our time.

They are basically antinuclear, making full use of the fears that are widespread within the population, and recently they have found new support in the so-called peace movement which holds that nuclear weapons must be banned at all cost, irrespective of other dangers to peace. Another motive is the wish to lead a simpler life, with better conservation of nature resources. Part of this is directed against the power of big business. Last not least, there is a strong opposition against bureaucracies, against the present form of government and the parliamentary system, or the domination of people by politics.

Scientists and engineers are caught in the middle of all this. Their responsibility must be to maintain a reasonable standard on the scientific and technical level. We have already seen that this is most difficult in view of the fact that technical proofs cannot be given in public discussions.

However, the most important characteristic of these movements is that they claim to have a monopoly over the protection of moral values. This is where they win over the scientists whose contribution usually leads mainly to a technical solution. The politicians, too, cannot easily resist such a movement once it has assumed a certain size, reaching five to ten per cent of the population as it does in Germany at present. Many decisions have already been influenced, or in many cases were prevented, by the existence of these movements and their influence on voters.

It is true that there is also strong resistance to such movements. It comes from all those, mainly in industry and the labour unions, who fear that the economic system cannot put up with such drastic change. Industry and government still get advice from 'classical' scientists. Here, with their claim to be technically competent and reliable, they are a stumbling block for the alternative movements.

This may explain why a concentrated effort has been made over quite a number of years to demolish the authority of the 'classical' scientists, and with considerable success as far as public discussion is concerned. This may have helped the discussion in some cases when the experts had become accustomed to taking a self-righteous attitude. But the more common method was to pretend that the scientists were all biased as well as being narrow specialists and, therefore, to demand that no committee should be without members officially biased in the opposite direction. This is now practised in many countries. What is more, it is demanded that the advice of experts, too, should be supplemented by the advice of anti-experts. Indeed, in Germany there exist some fifteen so-called 'eco-institutes' which are in part financed by ministries and even by industry. Their official aim is to counteract what they consider the inadequacies of the 'classical scientists' and institutions.

Our parliament has gone one step further than the government, installing a commission of inquiry on the future of atomic energy with an equal number of experts and anti-experts in which the fears expressed above have been realized. The result was a great mass of paper, minutes of meetings, opinions, counter-opinions, responses and counter-responses. Everything was discussed and nothing became clear. And what is worse, the experts were the best experts that could be found, there was nobody left who could have given a final opinion. Is this the end of technical advice to a parliament?

We have tried[5] to do something about this, by inventing a new sampling method by which from a great mass of evidence we select controversial issues and then test who is right, using a simple method; it seems that somebody who cheats on large matters cannot avoid cheating on small matters too, and here he can be caught. Or if a scientist speaks like a lawyer, concentrating on points which speak in favour of his cause and not mentioning points which he knows but which are unfavourable for his arguments, he is not credible as a scientist, and again usually he can be caught. In this way, we are able to measure the credibility of a person or a group. Here, I think, is another responsibility for the scientists.

I should mention, however, that the experts and the alternative movements need not be on opposite sides on all issues. One example is economic growth which the latter fight because they aim at a simpler life while the scientists feel that exponential growth cannot go on forever. Now that growth has come to a standstill in a number of countries, there might be a chance to promote proposals which aim at a prosperous economy without growth. I admit this sounds like an utopian idea today but maybe it will not remain so forever. The big problem is that we are now witnessing the greatest attack ever on our technical culture, world-wide or if I may say so half-world-wide, a movement which if it succeeds will work against progress in all its forms, wanted and unwanted progress, because one cannot divide the basis of knowledge. In our country, this movement has reached about ten per cent of the population, and we are not the only such country. We can ask why this is so and we can find or suspect many reasons, including those that have to do with the fact that we live in a time without war between the great powers but certainly not without efforts to gain advantages by other means. We can ask who is interested in preventing progress or in damaging our economy. This will not help much because we can never *prove* anything and it is easy to override such suspicions. But we can act in our own field. We see that there is a movement to discredit scientists. Here we can act because science, together with the humanities and the arts, is part of our culture, beyond our daily needs. The three must not be separated, they have the same roots. What they produce is much more than just useful facts and rules; they are an essential part of the dignity of the human race. In addition, and we may call this a byproduct, science is needed if we want to solve the practical problems of our time. We have every reason to defend science, as responsible scientists and citizens—citizens indeed of the world.

Notes and references

1. Jonas, Hans. *Das Prinzip Verantwortung.* Insel-Verlag. Frankfurt (1979).
2. One example may suffice: One of our politicians, Kurt Biedenkopf, recently said:

> The freedom of scientific research does not include freedom to decide on the application of scientific results. What science produces in pure research is an offer of options, of possibilities. Whether or not use is made of them is not for science to decide but for politics. These are questions to be answered by society in a political sense.

3. Rothman, S. and Lichter, S. R. 'The nuclear energy debate: scientists, the media, and the public'. *Public Opinion,* August-September pp. 47-52 (1982).
4. Frye, Northrop 'The bridge of language.' *Science* **212**, 121 (1981).
5. Maier-Leibnitz, H. 'Stichprobenverfahren zur Klärung wissenschaftlich-technischer Kontroversen.' *Naturwissenschaften,* **70**, 65-9 (1983).

6 *Animal experiment: benefit, responsibility and legislation*

W. D. M. PATON

Professor of Pharmacology and Fellow of Balliol College University of Oxford

I am sure that Professor Gowing is right in her view that scientists and politicians are not wholly distinct beings—and they are certainly linked by a common humanity. But it is worth stressing that however much passion or politics is involved in scientific activity, the ultimate objective is to be independent of it. Thomas Huxley's phrase about a beautiful hypothesis slain by an ugly fact is appropriate. Our ideal may be unattainable, but it is there: to arrive at conclusions that, independently of elegance, strength of advocacy, authority, or advantageousness, square with the evidence of experiment. So I hope I will be forgiven if, when more philosophical aspects of the subject come up, my primary test will be that of experience. For myself, that experience, briefly, is qualification in medicine, a year as a pathologist at a sanatorium, some years of human experiment in submarine physiology, and the rest in pharmacology, seeking to understand, improve, and teach about the remedies available in medicine, using all manner of techniques.

The meaning of the terms 'animal', 'experiment', 'fundamental' and 'basic'

First, I would like to ask a series of questions. Some of them raise rather general topics, but I think one must face them, because on the answers depend the objectives of the legislation. To begin with, '*What is included under the word 'animal'*'? The word is often restricted to non-human vertebrates, particularly the warm-blooded. But Leeuwenhoek[1] used it 300 years ago for creatures as small as protozoa and microbes. Albert Schweitzer[2], looking down his microscope, wondered if he had been justified in killing the micro-organisms there displayed. Neither the word, nor the defining characteristics of an animal are unambiguous. The 'scale of creation' is an old conception, and one can follow it along, seeking possible criteria. Begin with the inorganic world, and note that crystals grow. Next, take the viruses, which can reproduce in a suitable host; bacteria reproduce and swim. In the vegetable kingdom, flowers turn to the sun; the sensitive plant, *Mimosa pudica*, will fold its leaves at the noise of a handclap, and insectivorous plants like *Drosera*, as Charles Darwin showed, are exquisitely sensitive to chemicals. Invertebrates can be quite complicated: the jellyfish has a nervous system; and as to insects and cephalopods, if one watches an ant or a bee, or reads about the octopus, it is hard not to admit the words 'learning' or 'purpose' in some sense in describing their behaviour. The leech has receptors for painkilling opiate drugs. There is no need to elaborate on the range of behaviour of coldblooded and warmblooded vertebrates, ranging from frog and fish up to birds, primates and man. Scientific research involves the whole of this range. I do not think anyone has found an absolute, logically rigorous basis for creating a line of demarcation at any one point in the scale, whether one uses, as the criterion, complexity, reproduction, presence of a nervous system, signs of 'purposiveness', evidence of 'sentience', or of a pain-responsive system.

In the same way, of course, there is a continuous gradation between, say, lightness and darkness, with no logical divisions. Yet in practice we agree in being able to distinguish the two ideas, and to use the words 'light' and 'dark' effectively. I believe there is a similar general (though not universal) agreement that one can separate man from animals. This seems to me correct for one great reason: that there is no other species who remotely approaches him in his capacity to accumulate his experience, by the spoken, and (particularly) the written and the printed word. This accumulation is not like a coral reef—more and more of the same—but continuously changing. As a result sccessive generations of man can build on the accumulated achievements of their predecessors; and this creates a decisive, growing, and in effect qualitative difference. Man gains thereby both a unique responsibility and a unique value. It has recently been remarked that while some may term this view 'speciesism', others would call it 'humanism'.

I believe there is also agreement, although fewer people think about this, that for purposes of legislation one should draw a line somewhere in the region between vertebrates and invertebrates, as the Draft European Convention and the present British legislation does. That means that the tadpole, as a vertebrate, is included (I included 5000 tadpoles in my annual return to the Home Office a year or two ago). This line is drawn, if anything, conservatively; but its precise position is something which experience could modify.

These two empirical divisions characterize all proposed legislation: human experiment is to be handled separately, and the scale of creation beyond the vertebrates is excluded. An important point arises: if you argue that there is *not* sufficient reason to draw a line between man and animals, then there seems no adequate reason to draw a line anywhere else, for instance between vertebrates and invertebrates, or between 'animals' and 'plants', or even between the animate and the inanimate worlds. Indeed, there seems to me a grain of truth here: that man has a 'duty of care' to the whole of creation, because of his special capacities. But that does not solve the question of whether particular distinctions need to be made in practice. A natural cause for trying to make such distinctions can be identified, when close study is made of some creature, whether a primate, a dolphin, an octopus, a fish, or a bee. This always reveals capacities previously unrecognized, and is not only a legitimate, but an attractive activity. But if the knowledge gained is then used in comparisons between species to create some line of demarcation, that can be misleading. Such comparisons are only valid if the *same* attention, the *same* magnifying glass is used on other species too; and if you do that, then the strength of the proposed demarcation falls away. One can call it 'the fallacy of exaggerated attention', although one would welcome a pithier phrase.

One carries forward, therefore, two particular points. First, if one does not accept a working distinction between man and animals, then no other distinction is available, and one is left with an undifferentiated responsibility for all creation. One should take as much care of a blade of grass as of a cat or dog. But if one *does* accept the need for some working distinction, then that between man and vertebrates, and between vertebrates and the rest, seems reasonable to our experience.

Next, *what does one mean by the word 'experiment'*? It is vital to realize that we are discussing a process of discovering something that, before the experiment, was not known. This is the great objection to any legal provision which requires the beneficial outcome of an experiment to be precisely specified: it would not be an experiment if that were possible. The most you can attempt is to indicate a range of possible outcomes. Even that can only be done on the basis of pre-existing knowledge, and it is both common and important, that the unexpected occurs, in a most fruitful way. I believe all investigators, humble or elevated, would testify to this. There is one famous example: the late Sir Henry Dale was, as a young man, asked to study the uterine stimulant, ergot. It was a happy chance that different samples were contaminated (through bacterial action) with substances later identified as histamine, sympathetic amines, and acetylcholine[3]. The study of each of these led respectively in the succeeding decades to major advances in allergy, in knowledge of cardiovascular and other sympathetic control, and in the foundation of the idea of the neurotransmitter. There are many other examples, but let me mention two distinguished physiologists, both Presidents of our Royal Society, who have also given their experience: A. L. Hodgkin and A. F. Huxley; one cannot do better than to read their papers.[4]

My third question is: *What do 'fundamental' and 'applied' mean, as applied to research*? The distinction could be regarded as important in our present context, if fundamental research is regard as merely satisfying 'wanton curiosity', while it only 'applied' research, with concrete benefits in view, that could be readily justified. The common distinction, since the Rothschild report,[5] is of 'fundamental' simply adding to knowledge, and 'applied' being suitable for a detailed contract between investigator and customer. Personally, I believe it has been a great mistake that the third category of research, introduced in the Dainton report published simultaneously,[5] is little spoken of, namely 'strategic' research, that is, research of a fundamental type, but aimed at creating a baseline or framework from which applied work could grow. A very great deal of medical research, including all of my own, is of this type. What we should be clear about is this. It is correct, and useful, and I shall elaborate on this below, to think of the results of scientific research in terms of two categories—addition to knowledge, and practical benefit—but investigators do not always restrict their objectives to the one or the other; strategic research quite naturally aims at both, and one's objectives can easily change in midstream.

The significance of new knowledge

Let us then recognize the two great rewards from research—*new knowledge* and *practical benefit*—and now come nearer to the fundamental question: *What is the value of new knowledge won by animal experiment*? Sometimes it is dismissed as 'merely' satisfying curiosity. But its importance is very deep, and though it is hardly necessary to argue the case in a university, I would like to try one illustration. Suppose, for the sake of argument, that no animal experiments had been done in the past, but that scientific work on the inanimate kingdom had been unrestricted. We would now have a quite remarkably 'one-eyed' view of the natural world: on the one hand, a physics and chemistry that could be highly advanced, but on the other, a knowledge of the functioning of ourselves and of the animal kingdom inferior to that of the ancient Greeks. We would not know that our arteries in life contain blood pumped there by the heart, not air; or that activity in nerves causes muscles to contract; or even that the pinky-grey pulp inside our skulls was a central nervous system. Our outlook on the world, on ourselves, and on animals could only be profoundly different. The very knowledge of our relationship to the animal kingdom, which in part motivates the modern animal welfare movement, would be lacking. All this also applies to new knowledge won today. It seems to me, indeed, that we *dare* not cease trying to understand what we can of how our bodies and those of animals function; *we dare not let our knowledge of the inorganic world outstrip our knowledge of its inhabitants.*

What are the practical benefits from research?

This is a huge topic, and I must be very selective. It is sometimes argued that the improved health experienced by the industrial countries is due not to medical research (including animal experimental work), but, for example, to better nutrition, or to preventive medicine, or to changes in the diseases themselves. So, while we may point out that a number of discoveries requiring animal work, for instance the identification of pathogenic bacteria, the discovery of vitamins and vaccines, often form the *basis* of a preventive medicine procedure, we must be cautious about claiming too much. (It is worth stressing in passing that the various branches of medicine must not be regarded as in conflict, but as cooperating with and cross-fertilizing each other.) Nor, in looking for practical benefits achieved, can we simply list the remedies which some doctors, or some patients, believe effective; experience gained from controlled clinical trial, and of the placebo response, rules that out. Nor is it easy to obtain cogent statistics for remedies which improve *morbidity* without affecting mortality. Methods for measuring morbidity are essentially lacking, so that, for instance, the relief available to the sufferer from hay fever, the insomniac, the arthritic, the epileptic, the itching, the disfigured, is hard to quantify; so we must say little about that.

Below are a few specific examples that I believe to be cogent. Figures 1 and 2 show the effect of chemotherapy on puerperal fever and on lobar pneumonia; Figures 3 and 4 show the effects of chemotherapy on polio vaccination and on vaccination against measles (that old precursor of childhood bronchial pneumonia). Now we turn to two more chronic conditions: first, the effect of therapy for hypertension (Fig. 5)—initially with rather clumsy drugs, then steadily improved; and second, childhood leukaemia (Fig. 6)—a good example of how progress is usually made, not by a great 'breakthrough', but by steps, a few months more survival at a time: and, with the recent method of X-raying the skull to destroy leukaemic cells in the brain where they are beyond the reach of drugs (Fig. 7), a really remarkable final result, when one recalls the starting point. These figures show a considerable body of statistical evidence. Yet this hardly does justice to what seems so striking to one of my generation—that some medical appearances have just been abolished: for example lupus, i.e. tuberculosis of the skin (Fig. 8); Addison's disease, found to be due to destruction of the adrenal glands, where first an extract from the glands was found to prolong the life of an adrenal-deprived animal, then slowly the chemistry was solved, and the family of adrenocorticoid drugs so widely used today was created (Fig. 9); lastly, Hodgkin's disease—a sort of cousin of leukaemia—still not curable, but enormously palliated (Fig. 10). (Further material may be found in Dollery[6] and Paton[7]).

Figure 11 summarizes the data and also shows (note the geometrical scale) the growth and levelling off of animal experiment. When one reviews these, the following points emerge:

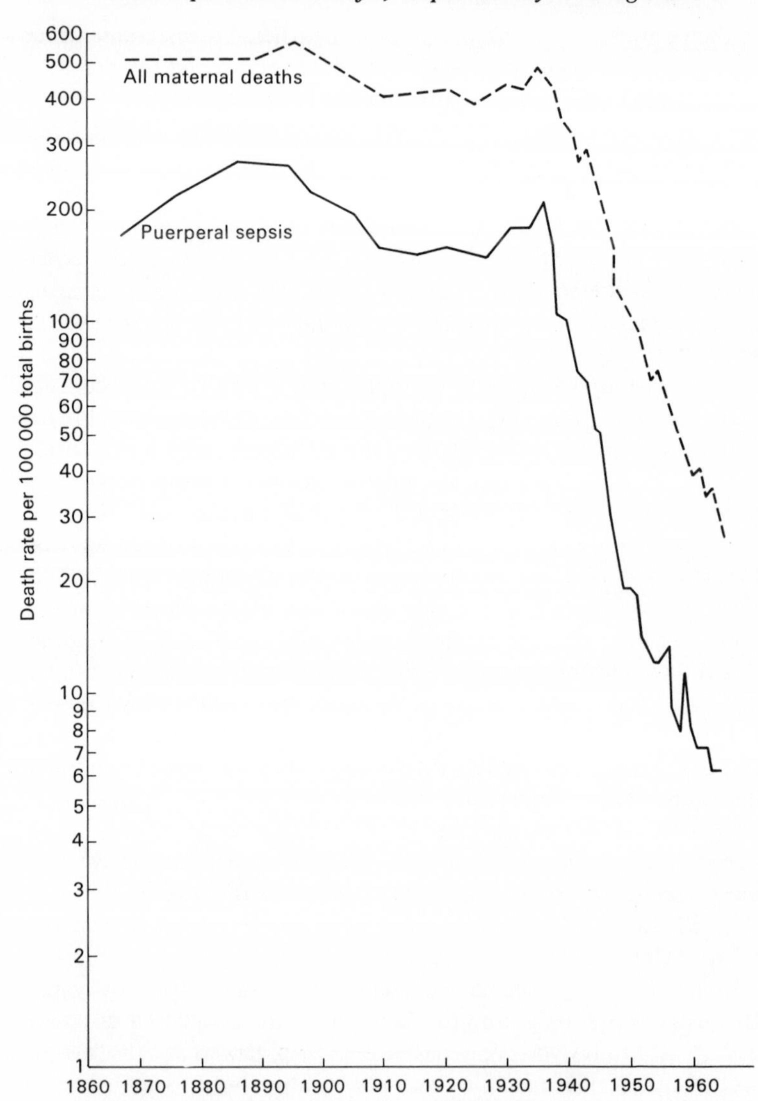

Fig. 1 Maternal mortality, death rate per 100 000 total births, 1860-1964, England and Wales. (*Source*: Derived from Registrar General's Decennial Supplement, England and Wales 1931. Part III. Registrar General's Statistical Review of England and Wales, various years.)
Note: (1) Ten years averages 1861-90, five-years averages 1891-1930, annual rates 1931-64. (2) Logarithmic scale.

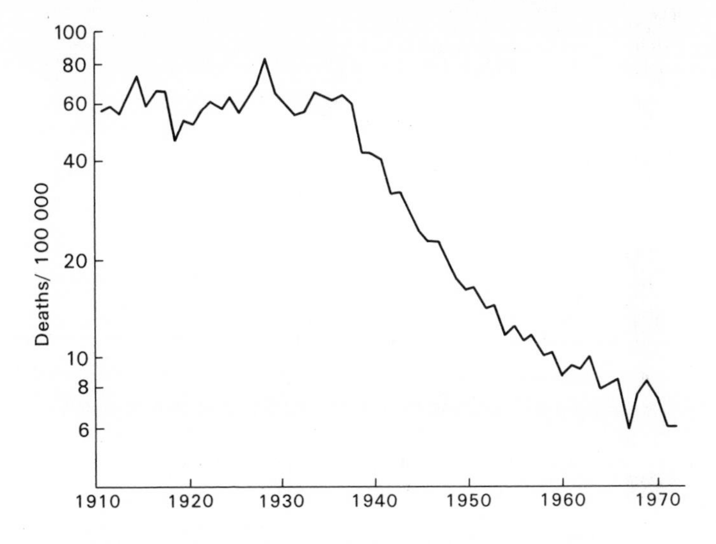

Fig. 2. Death-rate from lobar pneumonia in middle-aged men (45-64) in England and Wales, 1911-72

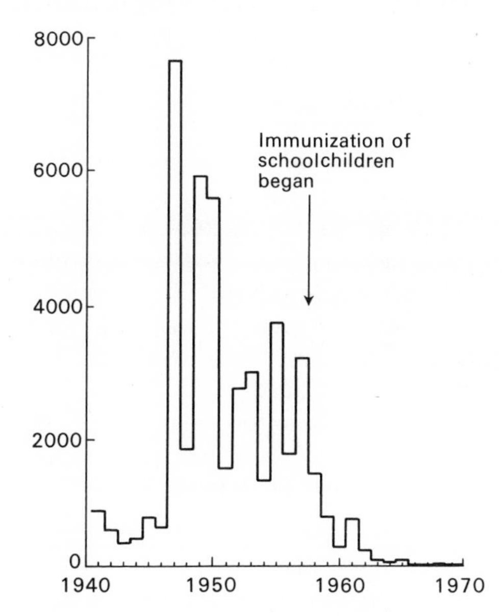

Fig. 3. Acute paralytic polio—annual notifications

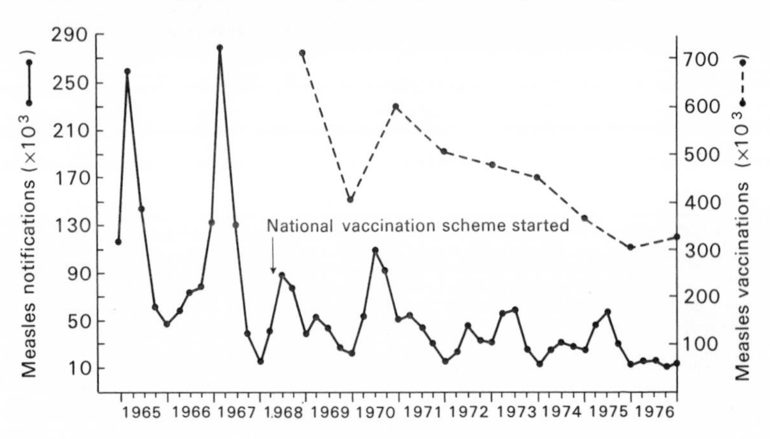

Fig. 4. Notifications of measles in quarterly periods October 1964 to December 1976 and measles vaccinations 1968-76, England and Wales.

(1) That there is often an interval of many decades between discovery and benefit.
(2) That each advance commonly (perhaps always) required some previous fundamental discovery, made with no purpose of immediate practical benefit; and that many of the practical benefits also produced new weapons for fundamental research: these points blur the 'fundamental/applied' dichotomy.
(3) That to be successful, all research must be 'fundamental' in one respect: that the investigator must be scrupulous in separating what he *wants* to find, from what he in fact observes.
(4) I mentioned earlier that human knowledge is cumulative. This is profoundly important, and means that a finite amount of experimental work yields an ongoing benefit. If one generation discovers an effective poliomyelitis vaccine, you can control the disease in future generations. This is especially significant in assessing the justification of animal experiment.
(5) The benefits are not restricted to man, but extend very fully to animals as well, both by research on animal disease and, because of the applicability of remedies for human disease, to many animal conditions.

Finally, I would suggest that, to weigh up the consequences of restricting or abolishing animal experiment *now*, one does the experiment of supposing such a restriction or abolition at some time in the past. Which of these advances should be lost or delayed?

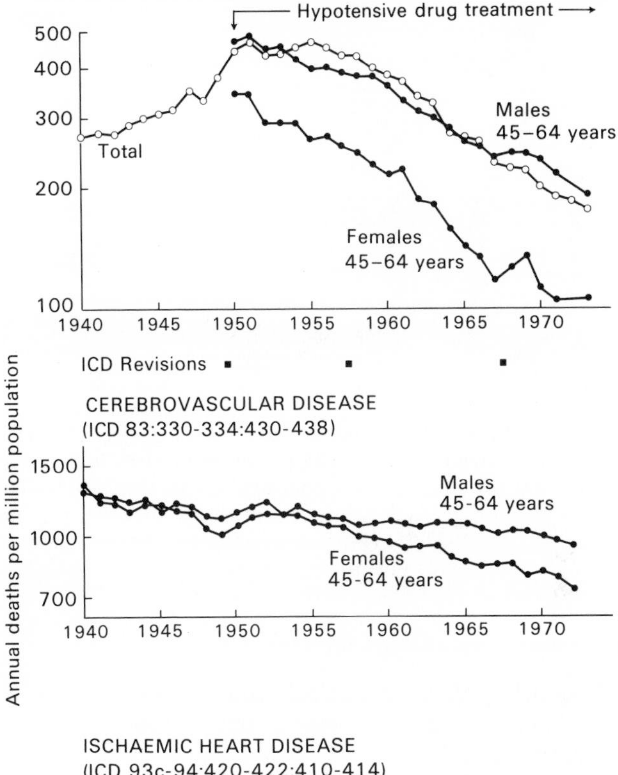

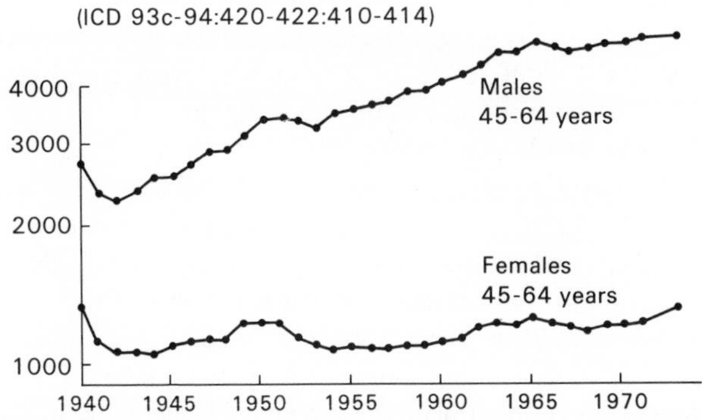

Fig. 5. Deaths from hypertensive, cerebrovascular, and ischaemic disease, 1940-73. (Data from Registrar-General's Statistical Reviews of England and Wales.) The International Classification of Diseases (ICD) was revised at the times indicated; the ICD categories are shown. Death rates, adjusted for change in the age and sex distribution in the population, have been extracted for males and females of age 45-64 years

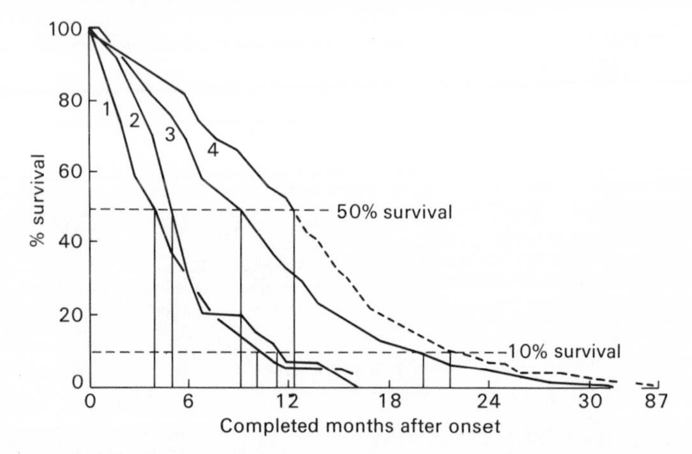

Fig. 6. Survival rates of patients under 15 years of age with acute leukaemia. (1) 218 untreated cases; (2) 34 cases treated with nitrogen mustard, 1946-48; (3) 160 cases treated with folic acid antagonists and/or steroids, 1948-52; (4) 184 cases treated with folic acid antagonists, steroids, mercaptopurine, azaserine, 1952-55. (From Burchenal and Krakoff, *Archives of internal Medicine,* **98**, 567 (1956))

Is it right to do experiments on animals?

It has been asked '*Is it right in principle to do to an animal what you would not do to a man*?' This is firstly a moral question. One could equally ask 'Is it right to do to a plant what you would not do to an animal?'; or 'Is it right to eat an animal if you would not eat a man?' or 'Is it right to do to a man what you could do on an animal?'. The answers do not rest on logic, but on a matter fundamental to the whole discussion; one's knowledge about, and view of, the relationship between man, animals and the rest of creation.

It is worth exploring for a moment the practical outcome of the view that man and animals are absolutely equivalent. That would not, of course, prevent animal experiment; it would mean that men and animals should be used according to which was most suitable for a particular study, and depending upon their relative cost and availability. Then one must ask whether, for instance, human subjects would be available, and that brings the further question of consent by men and animals. We know how to seek the consent of men, and a great deal of experiment is done on man. In fact, as we

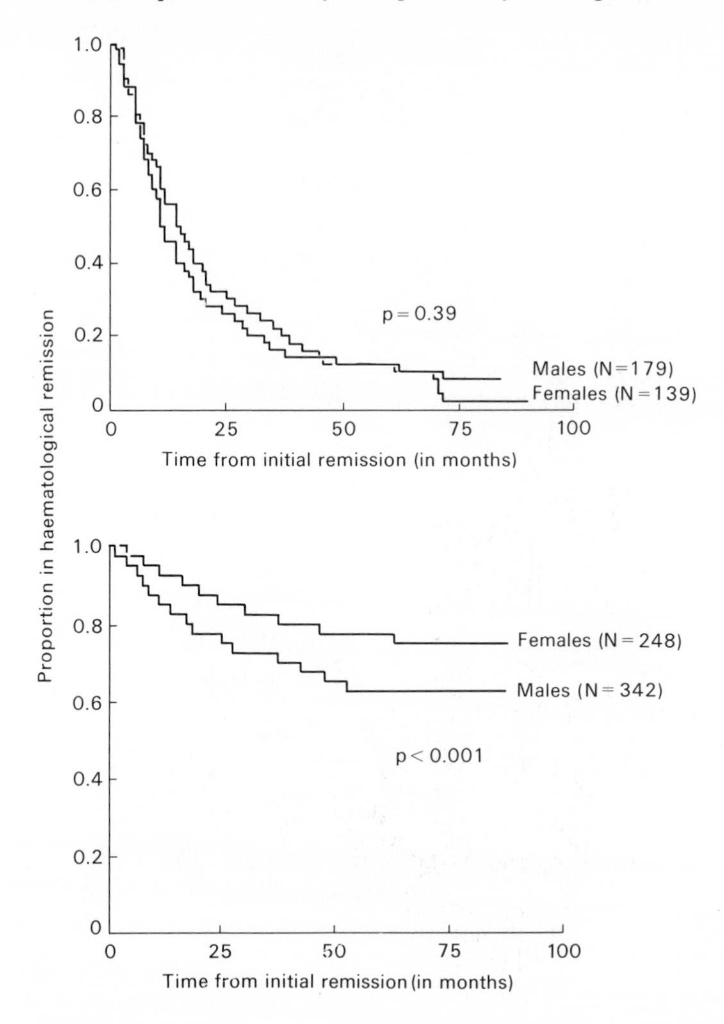

Fig. 7. Effect of therapy in childhood leukemia, measured by length of time for which leukemia cells were absent from the blood. *Above*, with chemotherapy only, shows remission in 50 per cent of cases for about 12 months (survival time will be appreciably longer). *Below*, combining chemotherapy with irradiation of the skull. Haematological remission for 60 per cent of over 100 months was obtained. It is believed that, today, 50 per cent of those under the age of 10 can be cured. (From Sather *et al.* (1981). Lancet **i**, 739.)

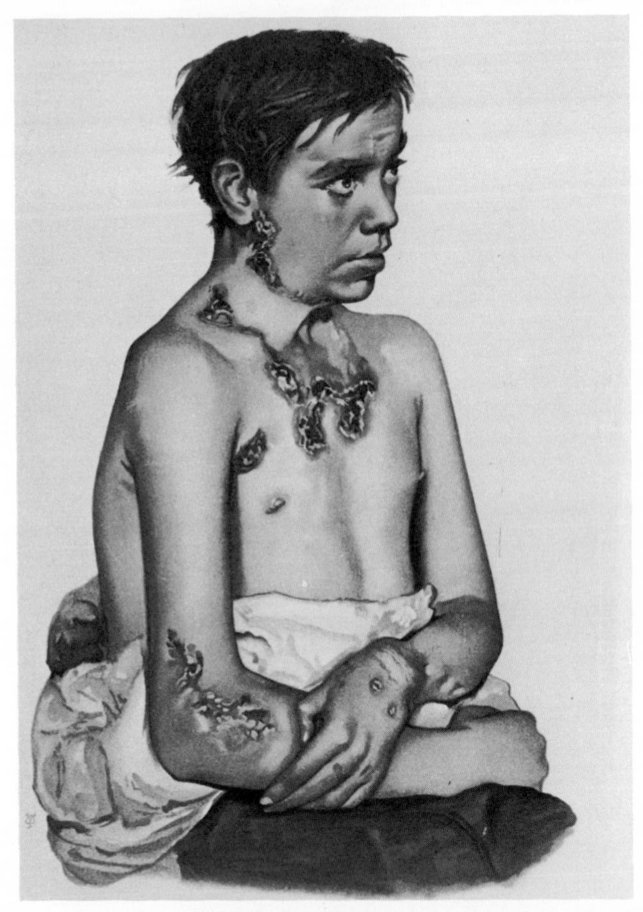

Fig. 8. A case of 'scrofula', tuberculosis of the skin (from Bramwell's *Clinical Atlas* vol. II, p. 4 (1893)).

shall see, I do not think there is scope for much more, when one looks at the literature in clinical pharmacology. But how do you obtain 'informed consent' from an animal? Indeed, do we obtain an animal's 'consent' to become domesticated or to become a pet? Again, the question of one's view of the relationship between men and animals is central. Those who argue for man-animal equivalence cannot answer on behalf of an animal, for then one would be accepting that men and animals are *not* equivalent.

The question 'Is it right?' may have yet another meaning: 'Is it *scientifically* right?'. This may be combined with the contention

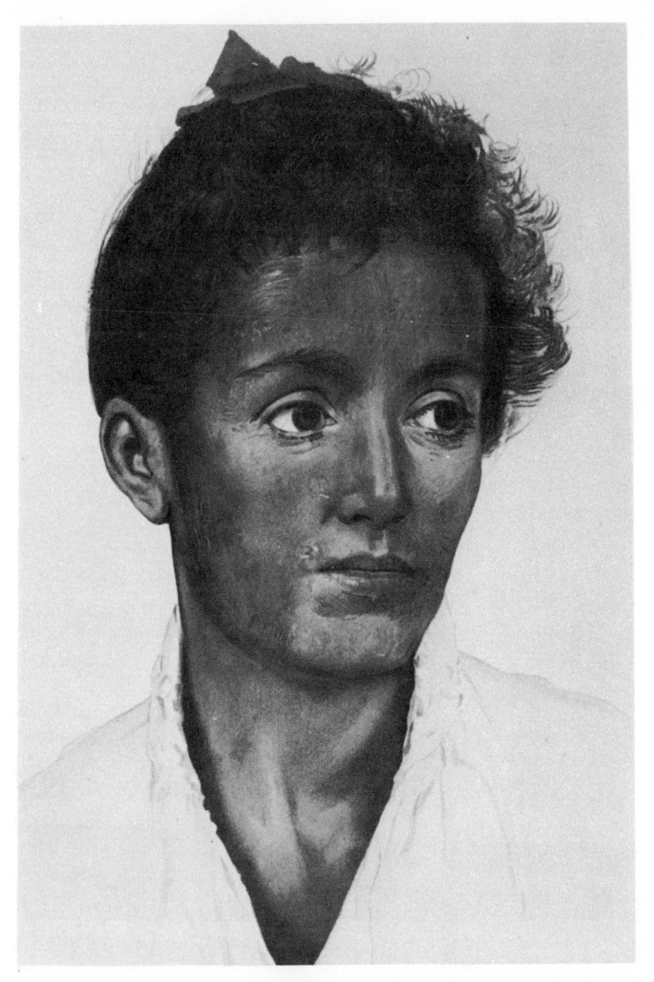

Fig. 9. A case of Addison's disease: a previously fair-complexioned lively woman aged 26 shows the characteristic pigmentation of skin and languor. (From Bramwell's *Clinical Atlas* vol. I., p. 72 (1892)).

that only work on humans can be relevant to human benefit, because of the biological differences between man and animals. On this view, all medical research should be done on man, and all veterinary research on animals, with neither able to benefit from the other. It seems to me illogical to claim simultaneously that man and animals are so alike that animals should be treated as men, and that

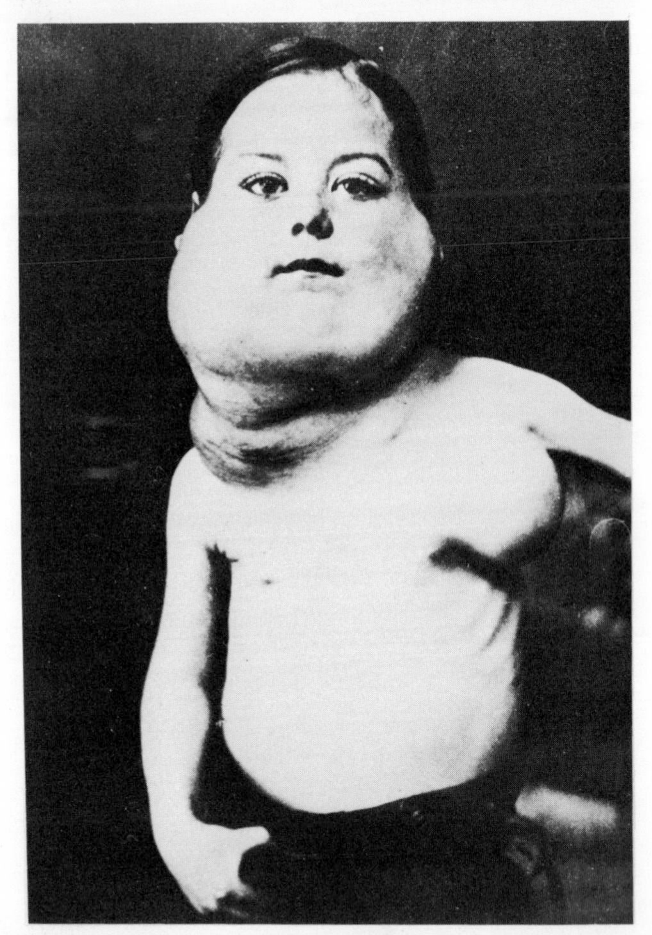

Fig. 10. A case of Hodgkin's disease, aged six (from Bramwell's *Clinical Atlas* vol. I., p. 94 (1892)).

man and animals are so unalike that research on them is irrelevant to human biology. However, the answer comes more from experience rather than from words or logic. As to the biological differences, it is certainly true that there *are* differences; perhaps the commonest in medical research are quantitative differences in metabolism. There are also striking similarities; for instance the guinea-pig and man are jointly unusual in both requiring vitamin C, and in being liable to allergy involving a particular mechanism. One must relaize that man varies within himself: e.g.

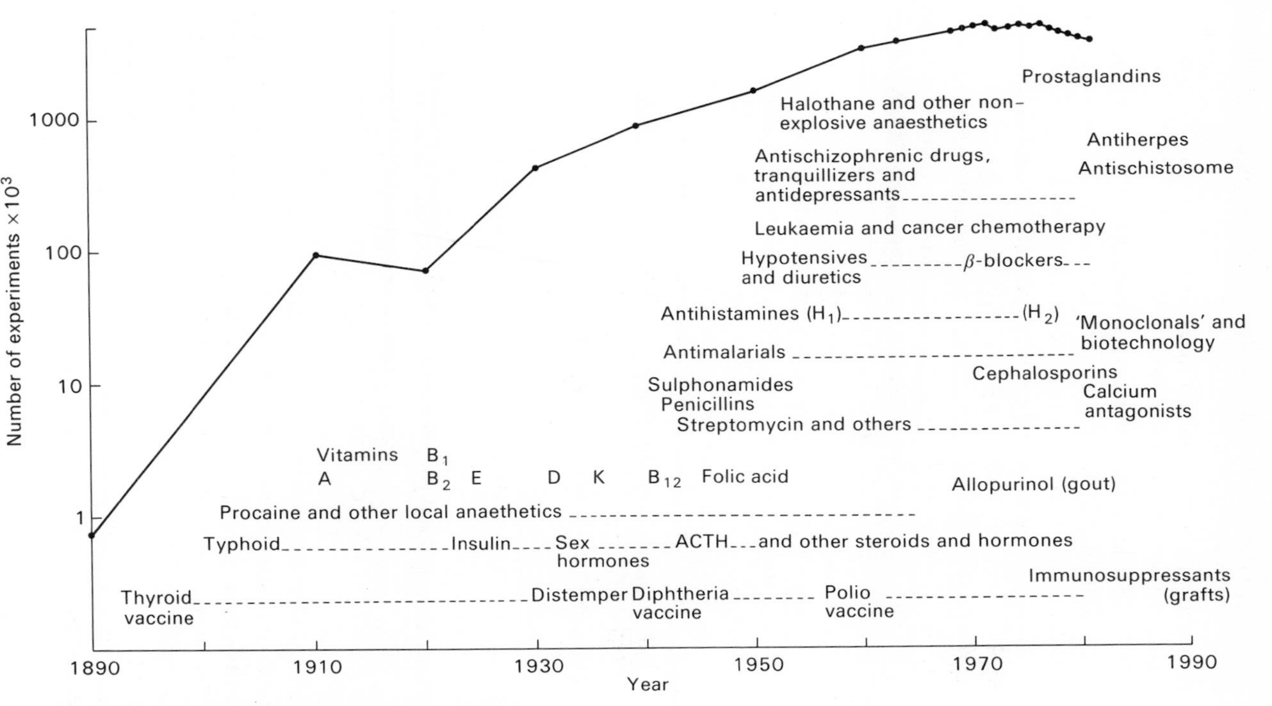

Fig. 11. Animal experiment and medical advance.

the slow acetylator, the low aldehyde dehydrogenase of the Japanese, the Mediterranean liability to sickle-cell anaemia, the latent myasthenic. Variation among animals is helpful, for one may find animals that reflect one of the genetic differences in man. But all these differences must not be overstated: the underlying patterns in man and in animals are *sufficiently alike to be immensely productive.* In my own field, if you consider *how* a drug works, it almost always has the same *type* of action in the animal as in the human body. Potency can vary, but even there, if one makes the proper correction for the dependence of metabolic rate on body weight, considerable regularity exists. Another striking fact is the enormous overlap between the pharmacopoeias used in both human and veterinary medicine (I have counted 80 drugs common to both). It is simply false to say that the biology of man and of animals is so different that knowledge about one is not applicable to the other: and, indeed, Fig. 11 (p. 105) illustrates that falsity in a striking way.

I conclude that the *moral* rightness of animal experiment, in principle, is a matter of personal and social decision: that the *practicality* of human experiment in its place, beyond the considerable amount already done, is small; but that the *scientific* rightness of animal experiment is well established.

What are the restraints on unnecessary animal experiment?

One may accept all this but still be deeply concerned that unnecessary experiments should not be done. One may then ask '*What are the restraints on unnecessary animal experiment*?' Two main criticisms are advanced: unnecessary repetition, and neglect of non-animal methods. First, *repetition*: some repetition is absolutely necessary: without it, a result can easily be dismissed as 'unconfirmed'. It is also essential, if others are to build on previous work, that they establish a genuine bridge to it, which involves, at the start at least, similar experiments. Often, too, control experiments are needed, and these may well appear repetitive. But one must recognize, too, the forces working *against* unnecessary repetitiveness: cost; waste of time; and the existence of many people (heads of departments, other scientists and grant-giving bodies) only too willing to point these out. The short answer is that the Littlewood Committee, in its study of British practice,[8] decided that there was little evidence to support the accusation of unnecessary repetition.

The scope of alternative methods

The second criticism is neglect of 'alternative methods': by this I mean methods which avoid doing experiments on whole animals, by using isolated tissues, or cells, or wholly non-animal methods. It is not always understood how far these other methods have been developed by animal experimenters themselves and how far they are used by them. There is a natural path that continually opens before the experimenter, as opportunities for deeper analysis on simpler systems arise in animal work. As a result, first an organ, then a tissue, then a cell, then a subcellular constituent may be used. This development depends on the previous animal work. It also needs to be carried back to work on the whole animal to establish its relevance. All the alternative methods I know of have traced this path.

There is, of course, a limit to what can be done with, for instance, tissue culture, despite the enormous advances that have taken place with such techniques. The difficulty is the complexity of the whole organism, which is worth explaining. There are a large number of active chemical substances normally present in the blood or the tissue fluid: nutrients, hormones, controlling proteins, electrolytes, a large number of biochemical transformations may occur. There are a large number of adaptive mechanisms: it is characteristic of biological organisms to react to changes in their environment, in a variety of ways. In each of these areas many other types of cell than those available in the culture are involved. Finally, especially in the brain, there are a vast number of connections made in the living animal with other cells, often at a considerable distance. In so far as a tissue culture fails to include the chemicals, the metabolic potential, the adaptive capacity, and the connectivity, for which other cells in the organism are essential, it moves steadily further and further away from the living organism. The use of cell culture methods thus needs prior information, mostly derived from animal or clinical experience, to identify where simplification is possible, and what supplementary factors must be provided. If successful, the advantages can be very great, not least because such work may become less laborious, because it can be automated, and (for production) because it can be scaled up. But there always remains the question of cell behaviour in its normal environment; and I think there is a good deal of truth in the prediction that too great a stress on tissue culture methods will simply increase the amount of animal work—by what is needed to establish the relevance of the methods to the whole organism.

The statement that experimenters neglect such methods always surprises me—not only for the reason just given, that they themselves developed the methods, but also because of the obvious evidence of extensive use. Table 1 lists some of the techniques familiar to me, with an indication of when they were first introduced. Going further, one can first compare biomedical journals now and in the past, and see both the growth of experiments on man, and the shift from the whole animal to isolated tissues, or from tissues to single cells, or from cells to subcellular constituents. For example I looked at the latest issues of the *British Journal of Clinical Pharmacology* and its sister the *British Journal of Pharmacology*. The former contained 20 papers, involving 124 experiments on normal human subjects and 99 on patients. The latter contained a similar number of papers: half of them on separated tissues, and all save two of the rest on animals anaesthetized and then killed. Of those remaining,

Table 1 *Methods of experiment not involving whole animals*

Isolated perfused organs (c. *1920*)	*Isolated tissue sample* (>*1900*)	*Isolated single cell* (c. *1908*)	*Subcellular constituents* (c. *1950*)
Liver	Striated muscle	Fat cells	Mitochondria
Muscle	Vas deferens	Liver cells	Microsomes
Heart	Seminal vesicle	Neurones	Synaptosomes
Lung	Electroplax	Glia	Actin/myosin
Adrenal	Smooth muscle	Muscle	Cell membranes
Intestine	iris, trachea	Smooth muscle	
Skin	bronchi, lung	Red cell	
Spleen	intestine, spleen	Leukocytes	
Kidney	uterus, bladder	Platelets	
	Salivary gland	Mast cells	
	Fat pads		
	Liver slice		

Structure-activity relationships in relation to	*Analogue or computer modelling* (c. 1870)
Curare (1868)	Drug metabolism and distribution
Nitrites (> 1900)	Receptor function
Barbiturates (> 1930)	Anaesthetic uptake
Sulphonamides (1940)	Decompression sickness
Sympathetic amines (1906)	Neuromuscular control
Anaesthetics (1900)	Nerve action potential
	Control of respiration

one involved injection of drugs to do with depressive disease and then killing the animal 30 minutes later, with the removal of tissue; the other was a test of antiepileptic drugs.

Second, in the United Kingdom at least, animal use has first levelled off, as Fig. 11 shows, and then declined; yet if you look at industrial or Medical Research Council reports, you find that in real terms biomedical research expenditure has steadily risen. Third, this corresponds with another UK statistic, that since 1970, the number of animal experiments done by each licensee has fallen by nearly 50 per cent—although the literature makes it clear that the number of experiments by research workers has certainly not declined. Finally, one may mention that in Britain it is the policy of the Medical Research Council and of the Home Office to encourage such alternatives.

That work with alternative methods, in those areas where they are applicable, should be supported, therefore, seems to me absolutely right. But that such work is *neglected*, I personally cannot accept.

To be candid, though, I must add a rider. If you ask 'where can I find the ideal assembly of tissue culture preparations, with cell growth limited to its natural bounds, ideally nourished, with *all* its interactions with other cells intact, *all* its receptors and synthetic material present?', the answer is obvious—the whole organism. Further, if you look back at medical discovery, one fact emerges, that the whole organism is a superb detector system. Only those who believe that we now understand most of how the whole body works can really believe that we can resolve our remaining ignorance on fragments. A more modest view recognizes that we have only just started plumbing the depths of complexity of biological function; so, one's strategy must include finding the best conditions where unexpected clues to what is still unsuspected may appear. I think there is a serious possibility, whatever one may think about toxicity testing, that for fundamental and strategic research, the objective of reducing whole animal experiment prematurely may merely prolong ignorance and unnecessary suffering.

The justification of animal experiment: striking the balance

My last question is a general one. 'To what extent are animal experiments in fundamental research *justified*?'. This is really a family of questions: 'Is this a well-designed experiment?' 'Is the precision sought in the experiment necessary?' 'How important is the knowledge to be gained?' 'Does it justify the suffering involved in the experimental procedure?' 'Is enough being done to maximize the knowledge and minimize the suffering?' Behind all these one can see three main issues: scientific value, amount of suffering if any, and striking the balance. First the *scientific value* (whether fundamental or applied) of the knowledge to be gained. This can only be reliably assessed or criticized by those who have familiarized themselves with the techniques and the scientific context of the research. Where does the required critique come from? It is liberally provided by one's colleagues, by learned societies and by grant-giving bodies, who commonly reject appplications for support because the expected scientific reward does not warrant the expenditure required. I think you would find that all those doing animal experiments would agree that this critique, by both peer judgement and economics, is a very powerful one.

The second issue is that of the *amount of suffering* if any, caused by the experimental procedures. How should that be assessed? Some people feel that it is not possible to make any statement about this. Others resort to a simple anthropomorphism. But a more usual view, ably expressed recently by Dr. Marion Dawkins,[8] is that it *is* possible to make rational judgements based on evidence; and she names a number of criteria, none sufficient by themselves, but jointly effective, and worth mentioning briefly: change in appearance or physical health; divergence from normal conditions in the wild; physiological evidence of stress; behavioural evidence; evidence from an animal's own choices; and extrapolation from our own experience. She argues cogently that, with each of these, a positive sign may not mean suffering, nor lack of such a sign mean lack of suffering,-but that combined they commonly allow judgement to be made. The task of assessment requires considerable experience; and this is an area where the veterinary profession, with its familiarity with both ill and healthy animals, can be of great use. One welcomes the recognition of this in proposed legislation.

The last issue is that of *balancing the scientific gain against the cost. How is that done*? In the end, as Lord Ashby has said, it is the conscience of the scientific worker which is the prime protection. It is sometimes claimed that the practice of animal experiment makes the experimenter callous and cold-blooded. Similar remarks are made about the medical profession; and the layman seeing a surgeon calmly dealing with a surgical emergency might feel that he, too, was unfeeling and cold-blooded. But one must distinguish between the superficial appearance of feeling and the actual exercise of professional skill and efficiency. The animal experimenter, too, must be professional and effective. In fact, my own experience is that animal work makes the investigator *more* sensitive to animal needs, as he learns about their behaviour, their physiology, and seemingly small but important matters about how to handle them.

But one must also remember that he does not work *in vacuo*; and just as when he is planning what type of experiment to do, or what scientific knowledge to seek, so here he is exposed to the influence of his colleagues (both immediately and through societies), of editorial boards of journals, and of grant-giving bodies—all of whom exert pressure, sometimes informally, sometimes by ethical codes or sometimes by ethical committees. In the United Kingdom, our Home Office inspectors play a role, which I believe to be of great value. Formally, their task is to supervise the administration of the law on animal experiment. But, in the course of this, the discussion by them with the investigator, particularly the young investigator, of proposed procedures, seems to me to exert a useful influence, just at the point where it is needed.

How can the general public be assured that a proper balance is struck?

I have just mentioned some of the reasons for assurance. But if you wish to go further, I think there is no escape from direct study of the extensive medical literature and the practice there displayed. Let me stress again that it is necessary to understand both *what* is done, and *why* it is done. I would like to illustrate this with a story. Among a number of experiments criticized as being obviously unjustified was one in which a building material, vermiculite, was injected into a rat's lungs. I, too, wondered how on earth this could be justified, and looked into it. This was the outcome: vermiculite is an insulating material, an alternative to asbestos; like asbestos, it owes its insulating property to its physical nature, very finely divided. It has been rather plausibly suggested that asbestos can cause cancer simply *because* its fibres are so fine—much smaller in diameter than the cells which it affects. Obviously vermiculite had to be tested. To make sure the test was effective, a matching quantity of asbestos was also injected into the pleural cavity of other animals. After a period during which weight was gained, with no overt ill-health in either group, the animals were killed. Microscopy showed that asbestos had begun its carcinogenic work: vermiculite was free of this action. In practical terms, then, I would regard this seemingly bizarre experiment as abundantly justified.

What lines should legislation follow?

The common criticism of our 1876 Act, largely on the simple ground that it is over 100 years old and therefore *must* be obsolete, has always seemed to me wrong. Our system remains a better one than that in use anywhere else in the world, and it is indeed a model. The reason, I believe, is that it was born after intensive and thorough debate—a debate in which most of the arguments were the same as those used today. I take the Act's principal features to be:

(1) the broadly drawn legislation with provision for regulation to meet changing circumstances;
(2) a procedure for approval of individuals, premises, and broad areas of work;
(3) an independent inspectoral system;
(4) the placing of responsibility explicitly on the person who actually does experiments; and
(5) the counting (and classification) of experiments actually done.

In Tables 2 and 3 I include some figures extracted from the return that may show how such information puts the issues in context. It is often claimed that the legislation is toothless, because there have been so few prosecutions under it. Others might regard that as evidence of success. But in any case, prosecution is not the real sanction; mere suspension or withdrawal of a licence is sufficient, with all this means for possible employment. For me, the main additional requirements are the inclusion of procedures involved in production (for example, of vaccines) rather than experiment, a modification of the system of licensee approval, and an explicit code of conduct of animal husbandry to take advantage of accumulated knowledge and experience. But, important as specific drafting is, I think the heart of the argument is in the sort of issues I have

Table 2 *United Kingdom Home Office return: numbers of experiments (1980)*

Categories			
Acute toxicity	484 849	Hazards and safety including	250 554
Chronic toxicity	251 395	environmental	40 892
Teratology	22 065	tobacco	1 868
Infection only	675 759	cosmetics and toiletries	31 304
Immunology only	589 071	pesticides and herbicides	57 604
Immunology and infection	429 224	household	13 832
		food additives	21 293
Neoplasia	208 700		
		Experiments on the eye	22 808

Table 3

United Kingdom Home Office returns of animal experiments

Year	*Total number of experiments (thousands)*	
1971	5607.4	100 per cent
1972	5327.1	
1973	5363.6	
1974	5561.2	
1975	5379.1	
1976	5474.7	
1977	5385.6	
1978	5195.4	
1979	4719.9	
1980	4579.5	
1981	4344.8	77 per cent

discussed, because these dictate the legislative aims. We may expect to hear a good deal more, however, about legislation.

Conclusion

The purpose of these lectures, I assume, is in the end not so much to assess what has happened in the past, but to look to the future. I would therefore like to end with some more forward-looking comments.

First, in my own lifetime, animal care has been, and is still being, revolutionized—partly by technical advances, partly by new knowledge, but also, I think, by continuous stepwise improvements by all those involved, scientists, animal house superintendents, and animal technicians. One result of the technical advance is a capacity to recognize earlier, and more sensitively, changes in animal well-being. This reduces animal suffering.

Second, on the scientific side, there has been a similar revolution in the techniques available, so that more precise, more economical and deeper questions can be asked. This, too, reduces animal suffering.

Third, recent advances in knowledge open up remarkable potentialities for further medical and biological advance. It is sometimes said that we know enough, or that we have enough drugs and treatments. But if we scan these new possibilities on the one hand, and on the other open our eyes to look at what remains to be done—in the developed countries, mental diseases (especially dementia, schizophrenia and depression), allergy, arthritis, neurological disease, congenital abnormality, stroke and cardiovascular disease, cancer; and in the developing world, along with all these, the whole field of tropical disease—then our prime duty is to see that the potential is not frustrated.

This brings me to my last issue, that of responsibility. It is not borne only by the research worker. I believe Lord Moulton's dictum of the last century remains true: 'Your duty is to take that line which produces the minimum total pain, and whether the pain is inflicted pain, or whether it is preventable pain that is not prevented, is in my opinion one and the same thing'. Today, for pain I woud say suffering, and Lord Moulton did not refer to the perpetuation of ignorance. But that is the task, with assessment of any inflicted suffering on the one hand, and on the other, assessment of preventable suffering—the magnitude of which we judge from both the historical record, and the present suffering we see around us. There have been recent television programmes on the subject: was it not irresponsible for these to have focused almost exclusively on inflicted suffering, and not at all on the historical record, or on present preventable suffering? Was it not irresponsible for the animal hooligans, mostly of the healthiest generation this country has ever seen as a result of the benefits of medical research, to harass individuals, breeders and firms? Was it not irresponsible for individuals to say that while, of course, they cannot condone violence, yet if something is not done soon, one must not be surprised at direct action—can condonation go further in a peaceful democracy? If animal experiment, essential as it is in medical and veterinary research, is unreasonably harassed or restricted now, then those that bring this about will carry a grave responsibility for the unnecessary ignorance and unnecessary human and animal suffering that will result in the future.

Notes and references

1. See Dobell, C. *Anthony v. Leeuwenhoek and his 'little animals'.* Staples Press, London (1932).

2. Schweitzer, A. *My life and thought.* p. 271. Allen and Unwin, London (1933).

3. Dale, H. H. *Adventures in physiology.* pp. x-xii. Pergamon Press, London (1953).

4. Hodgkin, A. L. *The pursuit of nature*, pp. 1-21 Cambridge University Press (1977); and Huxley, A. F. 'Discovery': accident or design.' *Proceedings of the Royal Society B,* **216**, 253-266 (1982).

5. HMSO Cmnd 4814 (1971).

6. Dollery, C. *The end of an age of optimism.* Nuffield Provincial Hospitals Trust. (1978).

7. See Paton, Sir William 'The evolution of therapeutics.' The Osler, Oration, 1978. *Journal of the Royal College of Physicians,* **13**, 74-83; (1979); Paton, W. D. M. 47th Stephen Paget Memorial Lecture, 1978. *Conquest*, February 1979, no. 169 (1979) and Paton, W. D. M. New York Academy of Sciences Workshop on the Role of Animals in Biomedical Research, April (1982); *Man and mouse.* Oxford University Press (1984).

8. Littlewood Committee. *Report of the Departmental Committee on experiments on animals*, Sections 248-256 HMSO Cmnd. 2641 (1965).

9. Dawkins, M. S. *Animal suffering.* Chapman and Hall, London (1980).

Index